B

Progress in Mathematics
Vol. 45

Edited by
J. Coates and
S. Helgason

Birkhäuser
Boston · Basel · Stuttgart

Jean-Michel Bismut

Large Deviations and the Malliavin Calculus

1984

Birkhäuser
Boston • Basel • Stuttgart

Author:

Jean-Michel Bismut
Département de Mathématique
Université de Paris-Sud
Bâtiment 425
91405 Orsay, France

Library of Congress Cataloging in Publication Data

Bismut, Jean-Michel.
  Large deviations and the Malliavin calculus.

  (Progress in mathematics ; vol. 45)
  Bibliography : p.
  1. Differential equations, Partial — Asymptotic
theory. 2. Manifolds (Mathematics) 3. Diffusion pro -
cesses. 4. Differential equations, Hypoelliptic.
I. Title. II. Series: Progress in mathematics (Boston,
Mass.) ; vol. 45.
QA614.9.B57   1984   515.3'53   84-3069
ISBN 0-8176-3220-4

CIP-Kurztitelaufnahme der Deutschen Bibliothek

Bismut, Jean-Michel:
Large deviations and the Malliavin calculus /
Jean-Michel Bismut. - Boston ; Basel ; Stuttgart :
Birkhäuser, 1984.
  (Progress in mathematics ; Vol. 45)
  ISBN 3-7643-3220-4   (Basel ...)
  ISBN 0-8176-3220-4 (Boston)

NE: GT

All rights reserved. No part of this publication may be reproduced, stored in a
retrieval system, or transmitted, in any form or by any means, electronic,
mechanical, photocopying, recording or otherwise, without prior permission of
the copyright owner.

© Birkhäuser Boston, Inc., 1984
ISBN 0-8176-3220-4
ISBN 3-7643-3220-4
Printed in USA
9 8 7 6 5 4 3 2 1

ABSTRACT

The purpose of this book is to use the Malliavin Calculus and large deviation techniques to study the asymptotics as $t \downarrow 0$ of the conditional probabilities of bridges associated with certain hypoelliptic diffusions. The program is fully completed in the elliptic case. In the hypoelliptic case, a deterministic Malliavin calculus is developed which exhibits the importance of the corresponding Malliavin covariance matrix in studying the curves of minimal action, and their relations to bicharacteristic curves. Two conjectures are formulated in the hypoelliptic case. A case study is done for the Heisenberg group.

AMS   Asymptotic behavior of solutions of PDE,
      Hypoelliptic equations and systems,
      Diffusion processes and stochastic analysis on manifolds,
      Gaussian processes, stochastic ordinary differential equations, Diffusion processes.

35B40, 35H05, 58G32, 60G15, 60 H 10, 60J60

Large deviations and the Malliavin calculus

Contents

Introduction 1

Chapter I : On the deterministic Malliavin calculus 17
- a) Notations and assumptions 18
- b) Some properties of the flows ϕ^h 20
- c) Conditions of invertibility of c_1^{h,x_0} 25
- d) Minimal action and the bicharacteristic flow 32
- e) The split of H and the differentials of I 43

Chapter II : Brownian motion on a Riemannian manifold and the Calculus of variations 54
- a) Notations and assumptions 55
- b) T^*M as a quotient of T^*N 57
- c) The Malliavin calculus of variations on a Riemannian manifold 60
- d) The conditional process 67
- e) The time reversed conditioned process 70
- f) An expression for $\dfrac{\mathrm{grad}_{x_0} p_t(x_0,y_0)}{p_t(x_0,y_0)}$ 78
- g) Semi-martingale property of the conditional process on [0,t] 86

Chapter III : Conditional diffusions and conditional flows : the basic estimates 90
- a) Reduction to the compact case 91
- b) The basic estimates 95
- c) Large deviations on flows 106

Chapter IV : Taylor expansion of the conditional probability : the elliptic case 117
- a) A first split of the Hilbert space H 120
- b) The split of the Brownian measure 124
- c) Expression of P_1 in terms of a standard Brownian bridge 127
- d) Large deviations for w^1 129
- e) A local change of variables 136
- f) An asymptotic expression for $p_t(x_0,y_0)$ 137
- g) The Jacobian of the exponential mapping in terms of a Brownian bridge 147

- h) A path integral proof of a result of Molchanov 149
- i) Taylor expansion of $p_t(x_0, y_0)$ 167
- j) A second split 174
- k) Expansion of $p_t(x_0, x_0)$ 177

Chapter V : The hypoelliptic case : Two conjectures
- a) Assumptions and notations 184
- b) Remarks on the semi-martingale property 186
- c) Two conjectures 188
- d) The two conjectures : the elliptic case 199
- e) A case study : the Heisenberg group 200

References 210

INTRODUCTION

The purpose of this book is to apply the methods of the Malliavin calculus to describe and solve some problems connected with the asymptotics (as $t \downarrow\downarrow 0$) of the semi group associated with an hypoelliptic diffusion.

1. A finite dimensional analogue

Before entering the heart of the matter, we will first describe a finite dimensional analogue of the infinite dimensional probabilistic model which we shall later use.

Namely let H denote a finite dimensional Euclidean space of dimension ℓ, which we endow with the canonical Gaussian measure dP. The general element of H is written w, and the Lebesgue measure on H is dw.

Let M be a C^∞ n-dimensional connected Riemannian manifold and let dx be the corresponding volume element in M. We assume that $\ell \geq n$. Let $\Psi(w)$ be a C^∞ mapping from H in M. We assume that if $\Psi'(w)$ is the derivative of $\Psi(w)$, $\Psi'(w)$ has maximal rank n. This is equivalent to saying that if $\Psi'^*(w)$ is the adjoint mapping of $\Psi'(w)$ (which maps $T^*_{\Psi(w)}M$ into H), if C(w) is defined by

(0.1) $\quad C(w) = (\Psi'\Psi'^*)(w)$

C(w) maps $T^*_{\Psi(w)}$ into $T_{\Psi(w)}M$ and defines a symmetric positive definite quadratic form on $T^*_{\Psi(w)}$, which we will call the Malliavin covariance matrix of the problem.

Under the previous assumptions, the law of $\Psi(w)$ is given by $p(x)dx$ where p is measurable. It was a key observation of Malliavin [46] - [47] - which is also interesting in the finite dimension case - that if $|C^{-1}(w)|$ is in all the L_p $(1 \leq p < +\infty)$, an integration by parts procedure combined with classical results on Fourier transform shows that $p(x) \in C^\infty(M)$.

We will now go one step further, i.e. evaluate $p(x)$. For $x \in M$, set

$$(0.2) \qquad K_x = \{w \in H \, ; \, \Psi(w) = x\}.$$

Since $\Psi'(w)$ has maximal rank, for every $x \in M$, K_x is a C^∞ manifold.

Let $*$ denote the usual duality operator on differential forms on H. Clearly

$$(0.3) \qquad dw = \frac{[\Psi'^*(w)dx] \wedge (*(\Psi'^*(w)dx))}{\|\Psi'^*(w)dx\|^2}$$

Now if $\Psi(w) = x$,

$$(0.4) \qquad d\sigma^x(w) = \frac{*(\Psi'^*(w)dx)}{\|\Psi'^*(w)dx\|}$$

is exactly the area element on K_x. Moreover it is easy to check that

$$\|\Psi'^*(w)dx\| = [\det C(w)]^{1/2}$$

(recall that M has a volume element, so that det C(w) is well defined).

The disintegration of dw on the manifolds K_x writes

(0.5) $$dw = \frac{dx\, d\sigma^x(w)}{[\det C(w)]^{1/2}}$$

and so

(0.6) $$dP(w) = \frac{e^{-\frac{|w|^2}{2}}}{[\sqrt{2\pi}]^{\ell}} \frac{dx\, d\sigma^x(w)}{[\det C(w)]^{1/2}}$$

It is then clear that

(0.7) $$p(x) = \frac{1}{(\sqrt{2\pi})^{\ell}} \int_{K_x} \frac{e^{-\frac{|w|^2}{2}}}{[\det C(w)]^{1/2}} d\sigma^x(w)$$

A regular disintegration of P with respect to $\Psi(w)$ is given for a.e. x by

(0.8) $$dP_x(w) = \frac{\frac{1}{(\sqrt{2\pi})^{\ell}} \frac{e^{-\frac{|w|^2}{2}} d\sigma^x(w)}{[\det C(w)]^{1/2}}}{p(x)}$$

If $p(x) < +\infty$ everywhere, (0.7) makes sense for every $x \in M$. At this stage, it should be pointed out that P is invariant under any one-to-one differentiable mapping τ which has determinant 1 and is such that $|\tau(w)|=|w|$. The r.h.s. of (0.6) is then invariant under a much larger class of transformation of H than the orthogonal group of H.

For $t > 0$, let $p_t(x)dx$ be the law of $\Psi(\sqrt{t}\, w)$. The same argument as in (0.7) shows that

$$(0.9) \qquad p_t(x) = \frac{1}{(\sqrt{2\pi\, t})^\ell} \int_{K_x} \frac{e^{-\frac{|w|^2}{2t}}}{[\det C(w)]^{1/2}} \, d\sigma^x(w).$$

Of course we will assume that (0.9) is $< +\infty$ for every $t > 0$, $x \in M$. We now want to study the asymptotics of $p_t(x)$ as $t \downarrow\downarrow 0$.

The r.h.s. of (0.9) can be studied using Laplace's method (see Azencott and al. [8]). Namely assume that $\lambda \in K_x$ is the unique element of K_x which minimizes the functional

$$(0.10) \qquad h \in K_x \to I(h) = \frac{|h|^2}{2}$$

The basic idea of Laplace's method is that the main contribution in the integral of the r.h.s. of (0.9) comes from any small neighborhood of λ.

Even in finite dimensions, there are two steps to estimate (0.9)

<u>Step n° 1</u> : We must prove that for any $\varepsilon > 0$, $k > 0$

$$(0.11) \qquad \frac{1}{(\sqrt{2\pi t})^\ell} \int_{K_x \cap |w-\lambda| > \varepsilon} \frac{e^{-\frac{|w|^2}{2t}}}{[\det C(w)]^{1/2}} \, d\sigma^x(w) = p_t(x) \circ (t^k)$$

Now (0.11) is certainly not a trivial estimate, except in the case where K_x is compact. If P_x^t denotes the conditional law of w given $\Psi(\sqrt{t}\, w) = x$, (0.11) is equivalent to proving that for any $k \in \mathbb{N}$

(0.11') $P_x^t(w \; ; \; |w-\lambda| > \varepsilon) = o(t^k)$

Step n° 2 : For one adequately chosen $\varepsilon > 0$, we must now evaluate

(0.12) $\dfrac{1}{(\sqrt{2\pi t})^\ell} \displaystyle\int_{K_x \cap |w-\lambda| \leq \varepsilon} \dfrac{e^{-\frac{|w|^2}{2t}}}{[\det C(w)]^{1/2}} \, d\sigma^x(w)$

Of course if I is defined on K_x by (0.10), $I'(\lambda) = 0$. Assume that $I''(\lambda)$ is positive - definite. Since $I'(\lambda) = 0$, $I''(\lambda)$ is a well-defined quadratic form on the Hilbert subspace H_1 of H

(0.13) $H_1 = T_\lambda K_x$

$\det I''(\lambda)$ is then well-defined on H_1. A classical argument shows that as $t \downarrow\downarrow 0$, (0.12) is equivalent to

(0.14) $\dfrac{e^{-\frac{|\lambda|^2}{2t}}}{(\sqrt{2\pi t})^n [\det I''(\lambda)]^{1/2} [\det C(\lambda)]^{1/2}}$

Using (0.11), we find that (0.14) is an equivalent for $p_t(x)$ as $t \downarrow\downarrow 0$.

To get a Taylor expansion of $p_t(x)$ of the type

(0.15) $p_t(x) = \dfrac{e^{-\frac{|\lambda|^2}{2t}}}{(\sqrt{2\pi t})^n} \, (q_0 + q_1 t + q_2 t^2 + \ldots q_k t^k + \ldots)$

we must proceed as in Erdelyi [79]. Of course the methods of [79] apply for the standard Lebesgue measure, so that we must now find a nice chart for K_x on a neighborhood of λ.

H_1 has been defined in (0.13). H_1 is also given by

$$H_1 = \{v \in H \; ; \; \Psi'(\lambda) v = 0\} \; .$$

H_2 denotes the orthogonal of H_1 in H. It consists of the $v \in H$ such that $q \in T_x^*M$ exists for which

(0.16) $v = \Psi'^*(\lambda) q$

and q in (0.16) is unique since $C(\lambda)$ is invertible. Now for $\epsilon > 0$ small enough, if $w^1 \in H_1$ is such that $|w^1| < \epsilon$, the equation on $q \in T_x^*M$, $|q| < \epsilon$

(0.17) $\Psi(w^1 + \lambda + \Psi'^*(\lambda)q) = x$

has one single solution $q(w^1,x)$. Set

$$v^2(w^1,x) = \Psi'^*(\lambda) \; q(w^1,x) \; .$$

Let P_1 be the standard gaussian measure on H_1. Of course $\lambda \in H_2$. An obvious change of variables shows that for $w \in K_x$ close enough to λ

(0.18) $$\frac{d\sigma^x(w)}{[\det C(w)]^{1/2}} = \frac{dw^1}{\det [\frac{\partial \Psi}{\partial v^2}(w^1 + \lambda + v^2(w^1,x))]}$$

or equivalently that

(0.19) $$\frac{d\sigma^x(w)}{[\det C(w)]^{1/2}} = \frac{[\det C(\lambda)]^{1/2} \, dw^1}{\det [\Psi'(w^1 + \lambda + v^2(w^1,x)) \, \Psi'^*(\lambda)]}$$

Set

(0.20) $$C(t,w^1,v_2) = \det [\Psi'(\sqrt{t}w^1 + \lambda + v^2)\Psi'^*(\lambda)]$$

Using (0.19), we find that the Taylor expansion (0.15) of $p_t(x)$ is obtained by doing the formal Taylor expansion of

(0.21) $$\frac{[\det C(\lambda)]^{1/2}}{(\sqrt{2\pi t})^n} \int_{H_1} \frac{\exp - \frac{|\lambda + v^2(\sqrt{t}\,w^1,x)|^2}{2t}}{[\det C(t,w^1,v^2(\sqrt{t}w^1,x))]} 1_{|\sqrt{t}w^1| \le \varepsilon} dP_1(w^1)$$

Of course ε is irrelevant in the final result. When expanding (0.21), the coefficients of t^k are terms of the type

(0.22) $$q_k = \int_{H_1} \frac{\exp\{-\frac{I''(\lambda)(w^1,w^1) - |w^1|^2}{2}\}}{[\det C(\lambda)]^{1/2}} d_{2k}(w^1) \, dP_1(w^1)$$

where the d_{2k} is a polynomial function of w^1 of degree 2k. Of course by (0.8), we see that the $d_{2k}(w^1)$ also define a Taylor expansion of the conditional law P_x^t.

2. The case of stochastic differential equations

Let $w = (w^1...w^m)$ be a m-dimensional Brownian motion. If $X_1(x),...,X_m(x)$ are m smooth vector fields on a manifold M, consider the stochastic differential equation

$$(0.23) \qquad dx = \sum_{i=1}^{m} X_i(x) \, dw^i$$

$$x(0) = x_0$$

Set

$$x_s = \Psi_s(w)$$

If $\mathcal{L}$ is the second order differential operator

$$(0.24) \qquad \mathcal{L} = \frac{\sum_{i=1}^{m} X_i^2}{2}$$

the law of $\Psi_1(w)$ is given by the kernel of $e^{\mathcal{L}}(x_0,.)$.

In [46]-[47], Malliavin showed that it was possible to define adequately the differentials $\Psi_1'(w)$. He introduced the covariance matrix of the problem

$$C(w) = \Psi_1'(w) \, \Psi_1'^*(w)$$

He proved that if $X_1...X_m$ verify the assumptions of Hörmander [35], then a.s. $C(w)$ is invertible, and that $|C^{-1}(w)|$ is in all the L_p ($1 \le p < +\infty$).

He then gave a probabilistic proof of the smoothness of the law of $\Psi_1(w)$ under Hörmander's assumptions. The Malliavin calculus has been the object of several new developments, in Bismut [11]-[16], Ikeda - Watanabe [36], Kusuoka - Stroock [45], Shigekawa [58], Stroock [61]-[63], Taniguchi [65]. For a survey, we refer to Bismut [14].

In this paper, we will try to extend the analysis done in (0.1) - (0.22) to the system (0.23), i.e. try to find the asymptotics of the law $p_t(x)$ of $\Psi_1(\sqrt{t}\,dw)$ as $t \downarrow\downarrow 0$ under Hörmander's assumptions. We obtain the analogue of (0.21), (0.22) in the elliptic case and formulate two conjectures in the hypoelliptic case.

There is indeed a first difficulty in the general hypoelliptic case. Namely by the results of Azencott [6] we know that large deviation techniques can be applied to diffusions like (0.23) and that as $t \downarrow\downarrow 0$ the trajectories of $\Psi_s(\sqrt{t}w)$ tend to "concentrate" along the deterministic paths of minimal horizontal action (the horizontal paths are the paths whose derivative is included in the vector space spanned by $X_1...X_m$). Now anybody familiar with the Malliavin calculus knows that the a.s. invertibility of $C(w)$ comes from the fact that there is a diffusion term in (0.23) (i.e. the paths of $\Psi_s(w)$ have unbounded variations and infinite energy). As $t \downarrow\downarrow 0$, the limit path will be in general deterministic, and smooth. Now if w is replaced by a smooth deterministic path λ, there is no reason that $C(\lambda)$ remains invertible. The analysis of (0.1) - (0.22) blows up a first time.

This is the reason why in section 1, we consider a standard differential equation of the type

$$(0.25) \quad dx = \sum_{i=1}^{m} X_i(x) \, \dot{h}^i \, ds$$

$$x(0) = x_0 \quad ,$$

whose solution is $\varphi_s^h(x_0)$. We define the deterministic covariance matrix

$$(0.26) \quad C_1^{h,x_0} = \frac{\partial \varphi_1^h}{\partial h}(x_0) \frac{\partial \varphi_1^h}{\partial h}^*(x_0)$$

and find under what conditions C_1^{h,x_0} is invertible.

This deterministic Malliavin calculus gives unexpected results. The invertibility of C_1^{h,x_0} is proved to be a property of the operator $\mathscr{L}$. C_1^{h,x_0} will be invertible for any $h \neq 0$, in the case where the set

$$(0.27) \quad \Sigma = \{(p,x) \in T^*M/\{0\}, \langle p, X_1(x) \rangle = \ldots = \langle p, X_m(x) \rangle = 0\}$$

of doubly characteristic points is a symplectic submanifold of T^*M (which is a weaker condition than ellipticity, and much stronger than Hörmander's conditions [35]). Now such a condition appears in Menikoff-Sjöstrand [50]-[51] where they study the heat equation for a class of pseudo-differential operators which includes (0.24).

The invertibility of C_1^{h,x_0} provides the critical link between the least action horizontal paths in M and the Hamiltonian flow in T^*M associated to the principal symbol of $\mathscr{L}$, so that the results of classical

Riemannian geometry extend under certain assumptions to hypoelliptic situations. It is even related to the smoothness of the minimal action function $\bar{E}$. On the Heisenberg group, C_1^{h,x_0} is in fact invertible for any $h \neq 0$; this gives a simple proof of a result of Gaveau [32]-[33] and Azencott [6] which was obtained by probabilistic methods !

A second difficulty for the application of (0.1) - (0.22) to hypoelliptic diffusions is the present lack of estimates of the type (0.11). For elliptic diffusions on a <u>complete</u> Riemannian manifold, the estimates of Varadhan [69]-[70] are exactly what is needed to prove the equivalent of (0.11) (in the paper we use the more precise estimates of Azencott [8] following Molchanov [54], but uniform estimates of Varadhan's type are quite enough). The fact that Varadhan's estimates blow up on non complete Riemannian manifolds (see Azencott [8]) is a clear indication that the estimates extending (0.11')(which are concentration results of the conditional law in any small tube containing the considered geodesic) are of a global nature.

This had led us to develop entirely the analogue of (0.1) - (0.22) for elliptic diffusions, leaving to a later paper the proof of the conjectures of Section 5 (for more general hypoelliptic diffusion) or the building of a counter example (which at this stage cannot be excluded).

In the elliptic case, in order to obtain geometrically invariant results, we use the descriptions of Malliavin [47], Eells and Elworthy [27]-[31] of the Brownian motion on a Riemannian manifold by means of its lift to the bundle of orthonormal frames. In Section 2, we establish some intermediary results on such Riemannian diffusions. In particular, we answer a query of Malliavin [48], by showing how the introduction of "orthogonal" variations of the driving Brownian motion permits us to obtain formulas of integration by parts which make disappear the unnecessary terms introduced by the lift to the bundle of frames. This new way of using the calculus of variations leads up to establish an exact representation of $\dfrac{\text{grad } p_t(x_0,y_0)}{p_t(x_0,y_0)}$ as the expectation for the Brownian bridge of a certain stochastic integral containing the Ricci tensor. This formula is very different from the formula of Elworthy - Truman [30] for $p_t(x_0,y_0)$.

In section 3, we do the hard analytical part, i.e. prove that Step 1 is indeed possible in the elliptic case. Since we lacked general large deviation results on stochastic flows for small time parameter, we need to prove more precise results on the drift of the considered conditional diffusion, so that we could prove that large deviation results are true for flows. We systematically use time reversal so that estimates are proved on $[0,t/2]$ using Ventcell-Freidlin [77], Donsker-Varadhan's theory of large deviations [23], [69], [70] and Varadhan's estimates, and extended to $[0,t]$ using time reversal. The fact that we work on the bundle of frames creates some difficulties in using time reversal (essentially because the lifted process cannot be adequately time reversed) which are solved using the theory of stochastic flows and general results on semi-martingales (Dellacherie - Meyer [18]).

In section 4, we prove (0.21) and (0.22) for the semi-group of an elliptic diffusion. H is now $L_2([0,1];R^n)$ so that P is the Gaussian cylindrical measure on H i.e. the Brownian measure. P_1 is the

Gaussian measure on a subspace H_1 of H, i.e. the law of a certain Brownian bridge w^1. The $d_{2k}(w^1)$ will be iterated stochastic integrals with respect to w^1, which happens to be a semi-martingale. It is a remarkable fact that the asymptotics of the conditional law are obtained through the interaction of the energy functional and of the corresponding Malliavin covariance matrix. The method appears to be an elaboration of the method of Levi sums (see Mc Kean-Singer [49]) where we can use the whole manifold of paths instead of just using the finite dimensional manifold M itself. It should be pointed out that even once the equivalent of (0.21) has been proved, the methods of Schilder [57] cannot be applied to expand (0.21). Indeed Schilder deals with truly continuous or differentiable functionals of w^1 which moreover have bounded differentials, while our functionals of w^1 have none of these properties. We must instead rely a careful estimates which are obtained by using in particular a result of Garsia-Rodemich-Rumsey [82].

Of course in the end, purely formal arguments leading to (0.21) can be justified, so that formula (0.21) can be "used". We indeed use (0.22) to compute the first term in the Minakshishundaram - Pleijel development of $p_t(x_0, x_0)$ [74] obtained in Mc Kean-Singer [49]. Of course (0.21), (0.22) have all the rotational invariance properties used in Seeley [76], and Atiyah-Bott-Patodi [5] for the construction of the parametrix of certain elliptic operators. Also the same procedure can be used for the Riemann-Kodaira operator □ and its associated semi-group. In the whole section, we use notations very close to (0.1) - (0.22), which may help the reader to follow the details.

In Section 5, we formulate the corresponding conjectures in the hypoelliptic case. One difficulty is that if e is the identity mapping on H_1, $I''(\lambda)$-e is not trace-class, but only Hilbert-Schmidt. However $\det[I''(\lambda)]$ is still well-defined, using the properties of the modified determinants of Hilbert-Schmidt operators (see Simon [59]). We verify that the natural extension of (0.14) coincides with the result of Gaveau [32] for the Heisenberg group.

In any case, it should be pointed out that we study the asymptotics of $p_t(x_0,y_0)$ as $t \downarrow\downarrow 0$ when x_0 and y_0 are not conjugate (this concept is well defined also in the hypoelliptic case by the results of section 1), so that the quadratic form $I''(\lambda)$ is positive definite. If $I''(\lambda)$ has a non zero kernel, it will be finite dimensional. Another split of the Hilbert space H is then necessary, and a detailed analysis of the corresponding singularity must be done. This reduces the problem to a finite dimensional problem, which can be analysed using Arnold's classification of singularities (Molchanov [54], Azencott and al [8], Duistermaat [25], [26]).

Our debt to existing litterature is considerable. First of all, Molchanov [54] had already indicated the possibility of expanding $p_t(x_0,y_0)$ in the elliptic case, while using non intrinsic methods. This Taylor expansion was also obtained in Kannai [40], using similar results for the wave equation. Our debt to mathematical physics is also obvious, since what we do is using Feynman integrals with real phase. In particular Elworthy and Truman [29] extending Truman [68] have found the expansion as $t \downarrow\downarrow 0$ of

$$(0.28) \quad \int_\Omega T_0(x_t) \exp - \frac{S_0(x_t)}{t} dP(w)$$

(where x_t is a Brownian motion on a Riemannian manifold) using variations on the horizontal lift very close to ours (they also have a potential V but this does not make much difference technically).

In [29], Elworthy and Truman also use large deviation techniques. The basic difference is that while they always work with the whole Brownian measure, we are forced to work on a Hilbert submanifold of H (on which the conditional laws are sorts of cylindrical measures), which makes appear the critical Malliavin covariance matrix (which somewhat reflects the curvature of this submanifold). More recently Davies and Truman [17] have expanded the trace of the heat equation operator for the Euclidean Laplacian plus a potential in terms of expectations of certain stochastic integrals with respect to the Euclidean Brownian bridge. We should also mention the pioneering work of De Witt - Morette [19] [20], De Witt - Morette - Maheswari - Nelson [21] . In [80], Fadeev - Slavnov introduce in quantum field theory a determinant whose effect is very close to Malliavin's covariance matrix. In a $-\Delta+V$ context (in the Euclidean space) Doss [24], Albeverio - Høegh - Krohn [2] and Albeverio - Blanchard - Høegh - Krohn [3] (who worked with Feynman integrals) are also relevant.

Recently Langouche and al. [81] have used path integrals to derive the semi-classical expansion for certain quantum mechanics propagators, which are associated to elliptic operators. The purpose of [81] is mostly computational. At a formal level, the computations of [81] have some similarity with what is done in our paper in the elliptic case. It seems to us that the Malliavin calculus is providing the mathematical tools which bridge the gap between the stochastic calculus and the elaborate machinery of path integrals on Riemannian manifolds as they are now used in the mathematical physics litterature.

Above all, I have a special debt to R. Azencott. He has announced, two years before this work was ever started, similar results in the elliptic case, which should be published [7]. Although I have not come to see his paper, he has told me that his methods were conceptually simpler, but that he does not work with intrinsic techniques. Apparently his methods do not obviously extend to the hypoelliptic case. We have had several stimulating (and exhausting) discussions.

J.M. Bony and J. Sjöstrand have given very useful informations and references on hypoelliptic operators. I took also much advantages of talks with B. Gaveau, and of several lectures and discussions with Prof. P. Malliavin.

The results in this paper have been announced in [83].

Note added in proof

Using the results contained in this book, we have been able to give a probabilistic proof of the index Theorem of Atiyah-Singer for classical elliptic complexes, and also to prove the corresponding Lefschetz fixed point formulas. These results have been announced in [84] and will appear in [85].

Also among references on the Taylor expansion of $p_t(x_0, y_0)$, we have omitted the interesting paper of Kifer [86], who proved that in the elliptic case, such an expansion can be constructed, this by using analytic and probabilistic methods. Finally let us point out a paper of Malliavin [87], where an abstract implicit function theorem is proved.

I - ON THE DETERMINISTIC MALLIAVIN CALCULUS.

As we have seen in the Introduction, we know from the results of Azencott [6] that in a loose sense, the conditional diffusions associated to the stochastic differential equation (0.23) tend to concentrate along the curves of least "horizontal" action.

In the case where the associated second-order differential operator (0.24) verifies Hörmander's assumptions [35], we know that the Malliavin covariance matrix is a.s. invertible [12]-[46]. The method of proof of this result involves very strongly the path properties of Brownian motion, so that for the limit deterministic curve, there is a priori no reason that the covariance matrix remains invertible.

In this section, we give conditions under which the Malliavin covariance matrix is still invertible on deterministic "horizontal" curves.

In a), we give the basic notations and assumptions. In b), the deterministic covariance matrix is defined. In c) conditions of invertibility of this matrix are given. It turns out that a much stronger assumption than Hörmander's implies this invertibility. This assumption is shown to be equivalent to the symplecticity of the manifold of doubly characteristic points of $\mathcal{L}$. Strangely enough, such a symplecticity condition already appeared in Menikoff-Sjöstrand [50]- [51]. It is verified for the Heisenberg group.

In d) we prove that if the Malliavin covariance matrix is invertible, the minimal action curves are projections of unbroken bicharacteristics. This fully answers a query of Gaveau [32]-[33], and Azencott [8], where such an equality was proved for the Heisenberg group using probabilistic arguments.

In e), we introduce the basic split of the Hilbert space $L_2([0,1]\ ;\ R^m)$ which will be constantly used in sections 3, 4, 5. The differentials of the action functional are computed along horizontal tied paths.

Only the results of e) are used in the other sections of the paper.

a) Notations and assumptions.

M is a connected C^∞ manifold of dimension n. In the sequel, we will assume that M is compact, or that $M = R^n$.

T^*M denotes the cotangent bundle of M. x will be the standard element of M, and (p,x) the standard element in T^*M. π is the canonical projection $(p,x) \in T^*M \to x \in M$.

pdx is the fundamental 1 form on T^*M. Its exterior differential S is the canonical symplectic form on T^*M, so that in local coordinates, S writes

(1.1) $\quad S = \sum_1^n dp_i \wedge dx^i$

(see Abraham-Marsden [1], Arnold [4]).

$X_1 \ldots X_m$ are m C^∞ vector fields on M. If $M = R^n$, we will assume that $X_1 \ldots X_m$ are bounded and that all their derivatives are also bounded.

$\mathcal{F}(M)$ is the set of C^∞ functions defined on M with values in M, which is endowed with the topology C_K^∞ of uniform convergence of the functions of $\mathcal{F}(M)$ and their derivatives on the compact subsets of M.

H is the Hilbert space $L_2([0,1] ; R^m)$. I is the function defined on H with values in R^+

(1.2) $\quad I(h) = \frac{1}{2} \int_0^1 |h|^2 \, ds$

For $h = (h^1 \ldots h^m) \in H$, consider the differential equation

(1.3) $\quad dx = (\sum_{i=1}^m X_i(x) h^i) dt$

$\quad x(0) = x_0$

If x^h is the unique solution of (1.3), set

(1.4) $\phi_t^h(x_0) = x_t^h$

It is clear that ϕ_t^h is a flow of C^∞ diffeomorphisms of M onto itself, which depends continuously on $t \in [0,1]$ for the topology C_K^∞.

If ψ is a diffeomorphism of M, ψ^* denotes its action on the tensors on M. In particular if $Y(x)$ is a vector field on M, $(\psi^{*-1}Y)(x)$ is the vector field

$$(\psi^{*-1}Y)(x) = [\frac{\partial \psi}{\partial x}(x)]^{-1} Y(\psi(x)).$$

b) Some properties of the flows $\phi_\cdot^h$.

We now give some obvious properties of continuous dependence of $\phi_\cdot^h$ on h, and introduce the deterministic Malliavin covariance matrix associated to $\phi_\cdot^h$.

Theorem 1.1 : If h^k converges weakly to h in H, then $\phi_\cdot^{h^k}$ converges to $\phi_\cdot^h$ uniformly on $[0,1]$ for the topology C_K^∞.
Moreover for any $x_0 \in M$, $t \in [0,1]$ the mapping $h \to \phi_t^h(x_0)$ is a C^∞ mapping from H in M.
In particular, if $h, h' \in H$ then

(1.5) $\qquad \dfrac{\partial \phi_t^h}{\partial h}(x_0)(h') = \phi_t^{h*} \displaystyle\int_0^t \sum_{i=1}^m (\phi_s^{h*-1} X_i)(x_0) h'^i ds$

Proof : The first part of the theorem is elementary using standard results on the continuous dependence on parameters of fixed points. The second part is a consequence of the implicit function theorem, of which (1.5) is an obvious consequence (also see [12] Theorem 2.1.)

□

We now define the Malliavin covariance matrix associated to ϕ^h ([12], [36], [46], [61]).

Definition 1.2 : For $x_0 \in M$, $t \in [0,1]$, $h \in H$, C_t^{h,x_0} is the linear mapping from $T_{x_0}^* M$ in $T_{x_0} M$ given by

$$(1.6) \quad p \in T_{x_0}^* M \to C_t^{h,x_0} p = \int_0^t \sum_{i=1}^m <(\phi_s^{h*-1} X_i)(x_0), p> (\phi_s^{h*-1} X_i)(x_0) ds$$

(in the sequel the summation sign $\sum_{i=1}^m$ will be omitted).

Of course C_t^{h,x_0} defines a symmetric non negative quadratic form on $T_{x_0}^* M$ given by

$$p \to <C_t^{h,x_0} p,p> = \int_0^t <(\phi_s^{h*-1} X_i)(x_0), p>^2 ds$$

Theorem 1.3 : Take $x_0 \in M$, $t \in [0,1]$, $h \in H$. Then C_t^{h,x_0} is invertible if and only if $\frac{\partial \phi_t^h}{\partial h}(x_0)$ (as a linear mapping from H in $T_{\phi_t^h(x_0)} M$) has maximal rank n.

Proof : Let $\widetilde{\frac{\partial \phi_t^h}{\partial h}}(x_0)$ be the adjoint mapping of $\frac{\partial \phi_t^h}{\partial h}(x_0)$, which is a linear mapping from $T^*_{\phi_t^h(x_0)} M$ into H. Clearly

$$\frac{\partial \phi_t^h}{\partial h}(x_0) \widetilde{\frac{\partial \phi_t^h}{\partial h}}(x_0) = \phi_t^{h*}(x_0) C_t^{h,x_0} \widetilde{\phi_t^{h*}}(x_0)$$

Now, the statement in the theorem is clearly equivalent to the fact that $\widetilde{\frac{\partial \phi_t^h}{\partial h}}(x_0)$ has a kernel reduced to $\{0\}$. The theorem follows. □.

Remark 1 : If $\mathcal{L}$ is the second order differential operator

$$\mathcal{L} = \frac{1}{2} \sum_1^m X_i^2$$

for one given $h \in H$, if $x_t = \phi_t^h(x_0)$, the invertibility of C_t^{h,x_0} only depends on the operator $\mathcal{L}$. To see this observe that if $\mathcal{L}$ can be written as

$$\mathcal{L} = \frac{1}{2} \sum_1^{m'} Y_j^2$$

(where the vector fields Y_j have the same properties as X_i), then we may express Y_j in the form

$$Y_j = \sum_{i=1}^m \lambda_j^i X_i$$

where $\lambda_j^i \in C^\infty(M)$. If $h' = (h'^1 \ldots h'^{m'}) \in L_2([0,1]; R^{m'})$, we can define the flow $\phi_\cdot^{\cdot h'}$ associated to the differential equation

$$dy = \sum_1^{m'} Y_j(y) h'^j ds$$

$$y(0) = y_0$$

Set

$$h^i(t) = \sum_{j=1}^{m'} \lambda_j^i(\phi_t^{\cdot h'}(x_0)) h'^j(t)$$

Clearly

$$\phi_t^h(x_0) = \phi_t^{\cdot h'}(x_0)$$

Moreover if $v' \in L_2([0,1]; R^{m'})$, $\frac{\partial \phi^{\cdot h'}}{\partial h'} t(x_0) v' = Z_t'$ is the solution of the differential equation

(1.7) $$dZ' = [\frac{\partial Y_j}{\partial x}(x_t) Z'h'^j + Y_j(x_t) v'^j] dt$$

$$Z'(0) = 0$$

where $x_t = \phi_t^h(x_0)$. (1.7) can be rewritten as

$$dZ' = [\frac{\partial X_i}{\partial x}(x_t) Z'h^i + X_i(x_t)[\frac{\partial \lambda_j^i(x_t)}{\partial x} Z'h'^j + \lambda_j^i(x_t) v'^j] dt$$

$$Z'(0) = 0$$

Setting

$$v^i_t = \frac{\partial \lambda^i_j}{\partial x}(x_t) Z' h'^j_t + \lambda^i_j(x_t) v'^j_t$$

we find that

(1.8) $\quad \frac{\partial \phi^{h'}_t}{\partial h'} t(x_0) v' = \frac{\partial \phi^h_t}{\partial h}(x_0) v$

(1.8) immediately implies that $\frac{\partial \phi^h}{\partial h} t(x_0)$ and $\frac{\partial \phi^{h'}}{\partial h'} t(x_0)$ have the same images. From Theorem 1.3, we see that C_t^{h,x_0} is invertible if and only if C_t^{h',x_0} (which is calculated for $Y_1 \ldots Y_m$,) is invertible. $\quad\square$

We now define :

Definition 1.4 : For $x, y \in M$ we define

(1.9) $\quad K^x_y = \{h \in H \,;\, \phi^h_1(x) = y\}$

Using time reversal, we see that there is a canonical one to one mapping from K^x_y into K^y_x given by $h \in K^x_y \to h' \in K^y_x$ where

(1.10) $\quad h'_t = - h_{1-t} \quad$ for $0 \le t \le 1$.

It should also be pointed out that if $h \in K^x_y$ is such that $C_1^{h,x}$ is invertible, then if h' is given by (1.10), $C_1^{h',y}$ is also invertible. This is obvious from (1.6).

We then have an easy consequence of Theorem 1.3.

__Theorem 1.5__ : If $x,y \in M$, if $h \in K_y^x$ is such that C_1^h is invertible, there exists a neighborhood $\mathcal{U}$ of h in H such that $K_y^x \cap \mathcal{U}$ is a submanifold of H. More precisely H splits into $H_1 \times H_2$, where $H_2 \sim R^n$, and a C^∞ diffeomorphism v of $\mathcal{U}$ on a neighborhood $\mathcal{U}'$ of $(0,0)$ in $H_1 \times H_2$ exists, such that

$$v(K_y^x \cap \mathcal{U}) = (H_1 \times \{0\}) \cap \mathcal{U}'.$$

__Proof__ : This is a consequence of Theorem 1.3 and of Theorem 1.3.8 in Klingenberg [41] □.

c) Conditions of invertibility of C_1^{h,x_o}.

We will now give conditions under which C_1^{h,x_o} is invertible. In particular we exhibit the relation between this invertibility and Hörmander's assumption [35] on the second order differential operator

(1.11) $\quad \mathcal{L} = \frac{1}{2} \sum_{i=1}^{m} X_i^2$

__Definition 1.6__ : $\mathcal{C}^o$ ([0,1] ; R^m) is the set of continuous functions w defined on [0,1] with values in R^m such that $w_o = 0$.

Of course $\mathcal{C}^o$ ([0,1]; R^m) is endowed with the topology of uniform convergence.

H is then densely embedded in $\mathscr{C}^o([0,1]\ ;\ R^m)$ through the mapping $u \in H \to e \in \mathscr{C}^o$ where

(1.12) $\quad e_t = \int_0^t u\,ds.$

Clearly the mapping $u \to e$ is continuous from H in $\mathscr{C}^o([0,1]\ ;\ R^m)$.

Let $\mathscr{C}^o$ be the topology induced by $\mathscr{C}^o([0,1]\ ;\ R^m)$ on H.

<u>Definition 1.7</u> : We will say $X_1,\ldots,X_m$ verify Hörmander's assumption H1 at x_0 if the vector subspace of $T_{x_0}M$ spanned by $X_1(x_0)\ldots X_m(x_0)$, and the Lie brackets at x_0 of $X_1\ldots X_m$ of any order is equal to $T_{x_0}M$.

We now have :

<u>Theorem 1.8</u> : If $X_1,\ldots,X_m$ verify H1 at x_0, then C_1^{h,x_0} is invertible on a $\mathscr{C}^o$ dense set in H.

<u>Proof</u> : Assume that a non empty open set 0 in $\mathscr{C}^o([0,1]\ ;\ R^m)$ exists such that for any $h \in H \cap 0$, C_1^{h,x_0} is non invertible.

Let P be the Wiener-measure on $\mathscr{C}^o([0,1]\ ;\ R^m)$. Consider the stochastic differential equation on $(\mathscr{C}^o([0,1]\ ;\ R^m);P)$

(1.13) $\quad dx' = \sum_1^m X_i(x').dw^i$

$\qquad x'(0) = x_0'$

and the associated flow of C^∞ diffeomorphisms $\psi_t(\omega, .)$; a.s., for any $x'_0 \in M$, $\psi_t(\omega, x'_0) = x'_t$ where x'_t is given by (1.13) (for the existence and properties of such flows see [10], [11], [31], [43]).

Now it is a basic result of Malliavin [12] - [46] that if C_t^{w,x_0} is the linear mapping $p \in T^*_{x_0} M \to C_t^{w,x_0} p = \int_0^t <(\psi_s^{*-1} X_i)(x_0), p> (\psi_s^{*-1} X_i)(x_0) ds$, then a.s., C_1^{w,x_0} is invertible.

For $k \in N$, $\ell \le 2^k$, $\frac{\ell-1}{2^k} < t \le \frac{\ell}{2^k}$, set

(1.14) $\quad h_t^k = 2^k [w_{\frac{\ell}{2^k}} - w_{\frac{\ell-1}{2^k}}]$

From Theorem I.4.1 in [10], we know that as $k \to +\infty$, $\phi_.^{h^k}$ converges to $\psi_.$ in probability for the topology C_K^∞ uniformly on [0,1]. It follows that as $k \to +\infty$, $C_1^{h^k,x_0}$ converges in probability to C_1^{w,x_0}.

Now since O is open and non empty, we know that $P(O) > 0$. A contradiction is then easily obtained.

□

The result of Theorem 1.8 is clearly unsufficient since as explained in the Introduction, we need to know if for one given $h \in H$, C_1^{h,x_0} is invertible.

We then consider the following assumption.

<u>Definition 1.9</u> : We say that $X_1,\ldots X_m$ verify assumption H2 at x_o if $m \leq n$, if $X_1(x_o) \ldots X_m(x_o)$ are independent, and if for any $\lambda = (\lambda^1\ldots\lambda^m) \in R^m$ such that $\lambda \neq 0$, if $Y = \sum_1^m \lambda^i X_i$, the vector subspace of $T_{x_o} M$ spanned by $X_1(x_o),\ldots,X_m(x_o)$, $[X_1,Y](x_o),\ldots[X_m,Y](x_o)$ is equal to $T_{x_o} M$.

It is easy to check that as assumption H1, H2 is in fact a property of the second order differential operator $\mathcal{L}$ defined in (1.11).

Clearly H2 is much stronger than H1.

We then have

<u>Theorem 1.10</u> : If $X_1(x_o) \ldots X_m(x_o)$ span $T_{x_o} M$ - i.e. if $\mathcal{L}$ is elliptic at x_o - for any $h \in H$, C_1^{h,x_o} is invertible.

If $X_1(x_o)\ldots X_m(x_o)$ does not span $T_{x_o} M$ - i.e. if $\mathcal{L}$ is not elliptic at x_o, C_1^{0,x_o} is not invertible.

If $X_1\ldots X_m$ verify assumption H2 at x_o, for any $h \in H$ such that $h \neq 0$, C_1^{h,x_o} is invertible.

<u>Proof</u> : The first part of the theorem is obvious by (1.6). Moreover, it is clear that

(1.15) $\quad <C_1^{o,x_o} p,p> = \sum_1^m <p,X_i(x_o)>^2$

The second part of the Theorem follows from (1.15).

We now prove the third part. If $h \in H$ is such that $h \neq 0$, we may consider the time change

$$\tau_t = \inf \{\tau \; ; \; \int_0^\tau |h| \, ds > t\} \; .$$

By considering the new flow ϕ_{τ_t}, it is easy to see that we may as well as assume that $h \neq 0$ a.e. on $[0,\varepsilon]$, for one given $\varepsilon > 0$.

Assume that $p \in T^*_{x_o} M$ is such that

(1.16) $\quad C_1^{h,x_o} p = 0,$

By (1.6), we see that

(1.17) $\quad <p, (\phi_t^{h*-1} X_i)(x_o)> = 0 \qquad 1 \leq i \leq m, \qquad 0 \leq t \leq 1$

If $Y(x)$ is any smooth vector field on M, it is easy to see that

(1.18) $\quad (\phi_t^{h*-1} Y)(x_o) = Y(x_o) + \int_0^t (\phi_s^{h*-1} [\sum_1^m X_j h^j, Y])(x_o) ds.$

From (1.17), (1.18), we find that for any $i(1 \le i \le m)$

(1.19) $<p,(\phi_t^{h*-1} [\sum_1^m X_j h^j, X_i])(x_0)> = 0$ a.e. on $[0,1]$.

Now h is a.e. $\neq 0$ on $[0,\varepsilon]$. Moreover for s small enough, $\phi_s^h(x_0)$ is such that H2 is verified at $\phi_s^h(x_0)$.

Using H2 and (1.17)-(1.19), it is clear that $p = 0$. C_1^{h,x_0} is then invertible □.

Remark 2 : The case where $\mathcal{L}$ is elliptic is the only case where, from the point of view of the calculus of variations, there is no difference between the stochastic case and the corresponding deterministic limit.

On the other hand, the second statement in Theorem 1.10 shows that in the evaluation of $p_t(x,y)$ for hypoelliptic diffusions as $t \to 0$, difficulties appear on the diagonal, i.e. at $y = x$. Of course for hypoelliptic left-invariant diffusions on nilpotent Lie groups, this fact immediately follows from mere homogeneity considerations on the corresponding graded Lie algebra (see Rotschild and Stein [56]) since $p_t(x,x)$ is easily seen to be equal to $\frac{C}{t^{Q/2}}$ (where Q is the weight of the Lie algebra) and Q is strictly larger than the dimension of the space. For the case of the Heisenberg group, see Gaveau [32]-[33], Azencott and al. [8] and Section 5.

Remark 3 : We may of course describe more complex conditions than H2 under which C_1^{h,x_0} is still invertible, and this by iterating equality (1.19) as in the stochastic Malliavin calculus [12]-[46]. This is left to the reader.

However it should be pointed out that the condition H2 already appears in the litterature of partial differential equations (see Menikoff and Sjöstrand [50]-[51]). To show this we now define :

Definition 1.11 : $\mathcal{H}(x,p)$ is the function defined on T^*M by

$$(1.20) \quad \mathcal{H}(x,p) = \frac{1}{2} \sum_{i=1}^{m} <p, X_i(x)>^2$$

Σ is the subset of $T^*M/_{\{0\}}$ given by

$$\Sigma = \{(x,p) \in T^*M/_{\{0\}} \; ; \; \mathcal{H}(x,p) = 0\}$$

Of course $\mathcal{H}$ is the principal symbol of $\mathcal{L}$, and Σ is the set of doubly characteristic points for $\mathcal{L}$ ([66] p. 331).

Recall that a submanifold N of T^*M is said to be symplectic if the restriction of the symplectic form S to N is non degenerate (so that N is itself a symplectic manifold).

We now have :

Proposition 1.12 : Under H2, there exists a neighborhood $\mathcal{V}$ of x_0 such that $\pi^{-1}\mathcal{V} \cap \Sigma$ is a symplectic submanifold of T^*M.

Proof : This is an easy exercise in symplectic geometry, which is left to the reader □.

Remark 4 : Menikoff and Sjöstrand [50]-[51] studied the heat kernel for a class of hypoelliptic operators under such a symplecticity assumption.

Remark 5 : Assumption H2 is trivially verified when M is the Heisenberg group of dimension 3, and $\mathcal{L}$ is the operator

$$(1.21) \quad \mathcal{L} = \frac{1}{2}(X_1^2 + X_2^2)$$

(where X_1, X_2 are the first two left invariant vector fields). This will prove to be interesting in the analysis of certain results of Gaveau [32]-[33] and Azencott and al. [8]. This fact will be used in Section 5.

d) Minimal action and the bicharacteristic flow.

Following Gaveau [32], [33], Azencott [6], [8], we now define the action associated to $X_1,\ldots,X_m$ and the corresponding Hamiltonian flow.

Recall that K_y^x has been defined in Definition 1.4.

$x_0 \in M$ is now fixed.

Definition 1.13 : The function $\bar{E}(y)$ is defined on M by

$$(1.22) \quad \bar{E}(y) = \inf_{h \in K_y^{x_0}} I(h) \quad \text{if } K_y^{x_0} \neq \emptyset$$

$$= +\infty \quad \text{if } K_y^{x_0} = \emptyset$$

We then have the following result.

Theorem 1.14 : $\bar{E}$ is a l.s.c. function with values in $R^+ \cup \{+\infty\}$. For every $y \in M$ such that $K_y^{x_0} \neq \emptyset$, there exists $h \in K_y^{x_0}$ such that

$$(1.23) \quad \bar{E}(y) = I(h)$$

If $X_1 \ldots X_m$ verify H1 at every $x \in M$, then for any $y \in M$, $K_y^{x_0}$ is non empty. In this case, the function $\bar{E}$ is finite and continuous on M.

Proof : The first part of the Theorem is contained in Azencott [6], [8] and is an easy consequence of the first part of Theorem 1.1.

Assume now that $X_1 \ldots X_m$ verify H1 at every $x \in M$. To prove that for any y, $K_y^{x_0}$ is non empty, we only need to prove that for any $x \in M$, there is a neighborhood $\mathcal{V}$ of x such that for any $y \in \mathcal{V}$, $K_y^x \neq \emptyset$. The result will then obviously follow from the connectedness of M.

Sjöstrand showed us how to establish such a local property. We will recall his argument in the simplest case. Namely assume that $x \in M$ is such that $X_1(x)$, $X_2(x)$ and $[X_1,X_2](x)$ span T_xM (so that $n \leq 3$). Consider the mapping ρ of R^3 in M given by

(1.24) If $t_3 \geq 0$, $\rho(t_1,t_2,t_3) = (\exp t_1 X_1 \exp t_2 X_2 \exp \sqrt{t_3} X_1$

$\exp \sqrt{t_3} X_2 \exp - \sqrt{t_3} X_1 \exp - \sqrt{t_3} X_2)(x)$

If $t_3 \leq 0$, $\rho(t_1,t_2,t_3) = (\exp t_1 X_1 \exp t_2 X_2 \exp - \sqrt{-t_3} X_1$

$\exp \sqrt{-t_3} X_2 \exp \sqrt{-t_3} X_1 \exp - \sqrt{-t_3} X_2)(x)$

Using the Campbell's Hausdorff formula, we know that

(1.25) If $t_3 \geq 0$ $(\exp \sqrt{t_3} X_1 \exp \sqrt{t_3} X_2 \exp - \sqrt{t_3} X_1 \exp - \sqrt{t_3} X_2)(x) =$

$\exp (t_3 [X_1,X_2] + o(t_3))(x)$

If $t_3 \leq 0$ $(\exp - \sqrt{-t_3} X_1 \exp \sqrt{-t_3} X_2 \exp \sqrt{-t_3} X_1 \exp - \sqrt{-t_3} X_2)(x) =$

$\exp (t_3 [X_1,X_2] + o(t_3))(x)$.

From (1.25), we see that ρ is C^1, and moreover it is a submersion at $(0,0,0)$. It follows that the image of the mapping ρ contains a neighborhood $\mathcal{V}$ of x. This method obviously extends to the case where more brackets are needed to span $T_x M$.

We now show that $\bar{E}$ is continuous. Take $x \in M$. Assume again that $X_1(x)$, $X_2(x)$ and $[X_1, X_2](x)$ span $T_x M$. Let $h \in K_x^{x_0}$ be such that

$$\bar{E}(x) = \frac{\int_0^1 |h|^2 \, ds}{2}$$

Consider the neighborhood $\mathcal{V}$ of x which has been constructed after (1.25). If $x_n \to x$, for n large enough, $x_n \in \mathcal{V}$. Using (1.24), (1.25), we may then connect x with x_n by a path of the form (1.24) so that

$$\rho(t_1^n, t_2^n, t_3^n) = x_n$$

and moreover we may assume that as $n \to +\infty$, $(t_1^n, t_2^n, t_3^n) \to 0$.

Set

(1.26) $\quad k_n = 1 - t_1^n - t_2^n - 4\sqrt{|t_3^n|}$

For n large enough $k_n > 0$. Consider the path $z^n(t)$ ($0 \leq t \leq 1$) which connects x_0 and x_n which will be briefly described as follows :

If $0 \leq t \leq k_n \quad z^n(t) = \phi_{t/k_n}^h (x_0)$.

If $k_n \leq t \leq 1$, $z^n(t)$ will be one of the two paths in (1.24). Namely if $t_3^n \geq 0$, z^n which is x at time k_n "follows" $-X_2$ during the time $\sqrt{t_3^n}$, then $-X_1$ during time $\sqrt{t_3^n}$ ... and finally X_2 during the time t_2^n and X_1 during the time t_1^n. If $t_3^n < 0$, the obvious change is made.

Clearly $h^n \in H$ exists such that $z^n(t) = \phi_t^{h^n}(x_o)$ and of course $h^n \in K_{x_n}^{x_o}$ so that

(1.27) $\quad \overline{E}(x_n) \leq \int_0^1 \frac{|h^n|^2}{2} ds$.

Since $t_1^n, t_2^n, t_3^n \to 0$ as $n \to +\infty$, it is easy to prove that $h^n \to h$ in H and so

(1.28) $\quad \overline{\lim} \, \overline{E}(x_n) \leq \int_0^1 \frac{|h|^2}{2} ds = \overline{E}(x)$

$\overline{E}$ is u.s.c. Since $\overline{E}$ is l.s.c., it is continuous. □ .

As in Gaveau [32], [33], Azencott [6], [8] we now define the associated Hamiltonian flow.

<u>Definition 1.15</u> : $Y(x,p)$ is the Hamiltonian vector field on T^*M associated to the Hamilton function $\mathcal{H}$ defined in (1.20).

In local coordinates, we have

$$(1.29) \quad Y(x,p) = \begin{pmatrix} \sum <p, X_i(x)> X_i(x) \\ -\sum <p, X_i(x)> \dfrac{\widetilde{\partial X_i}}{\partial x}(x) p \end{pmatrix}$$

Using the fact that $\mathcal{H}$ is invariant under Y, it can be easily proved that Y is a complete vector field on T^*M. Let ψ_t be the associated flow of diffeomorphisms on T^*M.

We also define the functions $H_1 \ldots H_m$ on T^*M by

$$H_i(x,p) = <p, X_i(x)> \, .$$

We then have the following elementary result.

<u>Proposition 1.16</u> : Take $(x_0, p_0) \in T^*M$. If $h \in H$ is defined by

$$(1.30) \quad h_t = (H_1(\psi_t(x_0,p_0)), \ldots, H_m(\psi_t(x_0,p_0)))$$

then for $0 \leq t \leq 1$

$$(1.31) \quad \psi_t(x_0,p_0) = (\phi_t^h(x_0), \phi_t^{h^*} p_0)$$

<u>Proof</u> : This is obvious from (1.29). □.

Note that since $\mathcal{H}$ is conserved by ψ, in (1.30), $|h_t|^2$ is constant and moreover

(1.32) $$\frac{\int_0^1 |h|^2 ds}{2} = \mathcal{H}(x_0, p_0).$$

The question then arises of knowing if the curves in M which minimize the action (1.22) are projections of integrals curves of the Hamiltonian flow ψ_t in T^*M, as is the case in classical Riemannian differential geometry (which corresponds to the case where $\mathcal{L}$ is elliptic). This question was raised by Azencott [8] for the Heisenberg group and only received a partial answer from the results of Gaveau [32] - [33].

Surprisingly enough, the study of the Malliavin covariance matrix C_1^{h,x_0} provides a reasonable answer to this question.

Namely we have :

Theorem 1.17 : Take $x \in M$. Assume that $K_x^{x_0}$ is non empty, and let $h \in K_x^{x_0}$ be such that

(1.33) $\bar{E}(x) = I(h)$.

Assume that C_1^{h,x_0} is invertible. Then there exists a unique $p_0 \in T^*_{x_0} M$ such that

(1.34) $\phi_t^h(x_0) = \pi \psi_t(x_0, p_0)$ $\qquad 0 \leq t \leq 1$.

In particular if H 2 is verified at x_o, for any $x \neq x_o$ such that $K_x^{x_o}$ is non empty, for any $h \in K_x^{x_o}$ such that (1.33) holds, a unique $p_o \in T_{x_o}^* M$ exists such that (1.34) is verified.

Proof : From Theorem 1.5, we know that since C_1^{h,x_o} is invertible, $K_x^{x_o}$ is a submanifold of H on a neighborhood of h. From (1.5) in Theorem 1.1, we find that the tangent space $T_h K_x^{x_o}$ consists of the $v \in H$ such that

$$(1.35) \qquad \int_0^1 (\phi_s^{h*-1} X_i)(x_o) v^i ds = 0.$$

Now the function I is differentiable on H. Since $K_x^{x_o}$ is locally a submanifold (on a neighborhood of h), the differential $dI(h)$ is 0 on $T_h K_x^{x_o}$. We then find that for any $v \in T_h K_x^{x_o}$,

$$(1.36) \qquad <dI(h),v> = \int_0^1 <h,v> = 0.$$

From (1.35)-(1.36), we find immediately that $p_o \in T_{x_o}^* M$ exists such that for $1 \leq i \leq m$

$$(1.37) \qquad h_t^i = <p_o, \phi_t^{h*-1} X_i> \quad \text{a.e. on } [0,1]$$

From Proposition 1.16, it is clear that

$$(1.38) \qquad \phi_t^h(x_o) = \pi \psi_t(x_o, p_o) \qquad 0 \leq t \leq 1.$$

Such a p_0 is unique because C_1^{h,x_0} is invertible.

If $h \in K_x^{x_0}$, if $x \neq x_0$, $h \neq 0$. The last part of the Theorem is a consequence of Theorem 1.10. □.

<u>Remark 6</u> : Assume that $M = R^n$. For $x \in M \neq x_0$, $\beta > 0$ we may consider the problem of minimizing on H

(1.39) $$\int_0^1 \frac{|h|^2}{2} ds + \frac{|\phi_1^h(x_0) - x|^2}{2\beta}$$

Clearly a minimum h^β exists. There is now no difficulty into applying the method of Theorem 1.17, so that $p_0^\beta \in T_{x_0}^* M$ exists for which (1.34) is verified (for h^β). Moreover

$$p(\psi_1(x_0, p_0^\beta)) = - \frac{\phi_1^{h^\beta}(x_0) - x}{\beta}$$

and finally

$$\int_0^1 \frac{|h^\beta|^2}{2} ds = \mathcal{H}(x_0, p_0^\beta).$$

If $K_x^{x_0}$ is non empty, it is easy to prove that as $\beta \to 0$, a subsequence of h^β - which we still note h^β - converges to $h \in K_x^{x_0}$ such that (1.33) holds, and so $\mathcal{H}(x_0, p_0^\beta)$ is uniformly bounded. However if $\mathcal{L}$ is not elliptic at x_0, there is a possibility that as $\beta \to 0$, $|p_0^\beta| \to +\infty$ because of the existence of doubly characteristic points. $\phi_1^h(x_0)$ may not be the projection of a bicharacteristic curve in T^*M.

Also observe that if $K_X^{x_0} \neq \phi$, as $\beta \to 0$, $\phi_1^{h^\beta}(x_0) \to x$, i.e. $\pi\psi_1(x_0, p_0^\beta) \to x$. From Theorem 1.14, we know that if $X_1...X_m$ verify H1 at every x, for any x, $K_X^{x_0} \neq \phi$. It is then obvious that if $X_1...X_m$ verify H1 at every x, $\pi\psi_1(x_0, T_{x_0}^* \mathbb{R}^n)$ is dense in $\mathbb{R}^n$. Of course all these results are also true if M is a C^∞ connected manifold.

This method (known as penalty method in control theory) shows a contrario that some assumption (like the one we did in Theorem 1.17) is needed for the result of Theorem 1.17 to hold.

In fact observe that

(1.40) $\quad h_s^{\beta, i} = <p_0^\beta, (\phi_s^{h^\beta *-1} X_i)(x_0)>$

and so

(1.41) $\quad \int_0^1 |h^\beta|^2 ds = <C_1^{h^\beta, x_0} p_0^\beta, p_0^\beta>$

If C_1^{h, x_0} is invertible, for β small enough, $|[C_1^{h^\beta, x_0}]^{-1}|$ will remain uniformly bounded, so that the sequence p_0^β has in fact a cluster point p_0.

In [32], Gaveau has shown that for certain nilpotent groups of rank 2, $p \to \pi\ \psi_1(x_0,p)$ is <u>not</u> onto, and still assumption H1 is verified, but of course H2 is not verified. In this case, from Theorem 1.14, for any x, $K_x^{x_0}$ is non empty and I has a minimum on $K_x^{x_0}$, which is not always the projection of a bicharacteristic curve.

Remark 7 : From Remark 5 and Theorem 1.17, we find that on the Heisenberg group G of dimension 3, when $\mathcal{L}$ is the operator (1.21), the result of Theorem 1.17 holds. This answers positively on the whole Heisenberg group a query of Azencott [8] : on the Heisenberg group, the curves of minimal action are indeed projections of unbroken bicharacteristics.

Remark 8 : If $x = x_0$, the result of Theorem 1.17 is of course trivially true under no special assumption.

e) The split of H and the differentials of I.

We now will explicitly compute (when this is possible) the differentials of any order of the action functional I on a minimal path.

In this subsection we do the following assumption:

H3 : $x \in M$ is such that $K_x^{x_o} \neq \emptyset$. Moreover $h \in K_x^{x_o}$ exists such that

(1.42) $\bar{E}(x) = I(h)$

and C_1^{h,x_o} is invertible.

At h, $K_x^{x_o}$ is then locally a submanifold of H, of codimension n.

We will first describe one "natural" parametrization of $K_x^{x_o}$ on a neighborhood of h.

<u>Definition 1.18</u> : H_1 is the set of $v \in H$ such that

(1.43) $\sum_{i=1}^{m} \int_0^1 (\phi_s^{h*-1} X_i)(x_o) v^i ds = 0$.

H_2 is the finite dimensional subspace of H which is the image of $T_{x_o}^* M$ by the linear mapping ρ

(1.44) $p \in T_{x_o}^* M \to \rho(p) = (<p, \phi_t^{h*-1} X_1(x_o)>, \ldots <p, (\phi_t^{h*-1} X_m)(x_o)>)$

Observe that since C_1^{h,x_0} is invertible, Ker $\rho = \{0\}$, so that topologically, H_2 can be identified to $T^*_{x_0}M$.

We then have the obvious:

<u>Theorem 1.19</u> : H_1 and H_2 are orthogonal subspaces of H and moreover

(1.45) $\qquad H = H_1 \oplus H_2$

If $v \in H$, the orthogonal projection $P_2 v$ of v on H_2 is given by

$$(P_2 v)_t = \rho(p(v))$$

where

(1.46) $\qquad p(v) = [C_1^{h,x_0}]^{-1} \int_0^1 (\sum_{i=1}^m (\phi_s^{h*-1} X_i)(x_0) v^i) ds$

<u>Proof</u> : The proof is elementary and is left to the reader. □.

We will now parametrize $K_X^{x_0}$ on a neighborhood of h using H_1.

<u>Theorem 1.20</u> : There exist $\varepsilon > 0$ such that if $v_1 \in H_1$ is such that $\|v_1\| < \varepsilon$, then there is one unique $v_2 \in H_2$ such that $\|v_2\| < \varepsilon$ and moreover

(1.47) $\qquad \phi_1^{h+v_1+v_2}(x_0) = x.$

If $v_2 = k(v_1)$, the mapping k is a C^∞ mapping from $\{v_1 \in H_1 ; \|v_1\| < \varepsilon\}$ into H_2.

Proof : From (1.5) in Theorem 1.1, we know that if $v_2 = \rho(p) \in H_2$, then

(1.48) $\qquad \dfrac{\partial \phi_1^h}{\partial h}(x_o) v_2 = \phi_1^{h*}(x_o) \, C_1^{h,x_o} \, p$

It immediately follows that $\dfrac{\partial \phi_1^h}{\partial h}(x_o)$, when restricted to H_2, is invertible as a linear mapping from H_2 into $T_x M$. The result is now obvious by the implicit function theorem. $\square$.

Of course $k(0) = 0$. We now show how to compute the differentials $\dfrac{\partial^m k}{\partial v_1^m}$.

We will use the following elementary result.

Proposition 1.21 : If $Y(x)$ is a smooth vector field on M, for any t such that $0 \le t \le 1$, and any $v \in H$

(1.49) $\qquad \dfrac{\partial}{\partial h}(\phi_t^{h*-1} Y)(x_o)(v) = [\int_0^t (\phi_s^{h*-1} X_i)(x_o) v^i ds, (\phi_t^{h*-1} Y)(x_o)]$

Proof : The proof is elementary and is given in [12] - Theorem 4.2. $\square$.

By the implicit function Theorem, to compute $\dfrac{\partial^m k}{\partial v_1^m}$, we differentiate

(1.47) as many times as needed and use Theorem 1.19 and Proposition 1.21 to do the computations explicitly.

The first two derivatives of k at 0 are now explicitly given.

Theorem 1.22 : We have

(1.50) $\dfrac{\partial k}{\partial v_1}(0) = 0$

Moreover if B is the symmetric bilinear form defined on $H_1 \times H_1$ with values in $T_{x_0} M$

(1.51) $(v_1', v_1'') \in H_1 \times H_1 \to B(v_1', v_1'') = \displaystyle\int_{0 \leq s \leq t \leq 1} [(\phi_s^{h*-1} X_i)(x_0) v_1'^i(s),$

$(\phi_t^{h*-1} X_j)(x_0) v_1''^j(t)] \, dsdt$

then

(1.52) $\dfrac{\partial^2 k}{\partial v_1^2}(0)(v_1', v_1'') = - \rho [C_1^{h,x_0}]^{-1} B(v_1', v_1'')$

Proof : We differentiate (1.47) i.e. we write that

(1.53) $\left(\dfrac{\partial \phi_1^{h+v_1+v_2}}{\partial h}\right)(v_1' + \dfrac{\partial k}{\partial v}(v_1) v_1') = 0$

From (1.5) in Theorem 1.1, if $v_2' = \frac{\partial k}{\partial v} v_1'$, we get

$$(1.54) \quad \int_0^1 (\phi_s^{h+v_1+v_2 \ast -1} X_i)(x_0)(v_1'^i + v_2'^i) ds = 0.$$

At $v_1 = 0$, $v_2 = 0$; using the definitions of H_1, we find that $v_2' = 0$, so that (1.50) holds.

Using (1.49), we differentiate (1.54). If $\frac{\partial^2 k}{\partial v_1^2}(v_1)(v_1', v_1'') = r$, we find that

$$(1.55) \quad \int_{0 \leq s \leq t \leq 1} [(\phi_s^{h+v_1+v_2 \ast -1} X_i)(x_0)(v_1''^i + v_2''^i)(s), (\phi_t^{h+v_1+v_2 \ast -1} X_j)(x_0)(v_1'^j + v_2'^j)(t)]$$

$$ds\, dt + \int_0^1 (\phi_s^{h+v_1+v_2 \ast -1} X_i)(x_0) r^i ds = 0$$

We do $v_1 = 0$, $v_2 = 0$ in (1.55). Since $v_2' = v_2'' = 0$, using Theorem 1.19, (1.52) is obvious. Note that the symmetry of B can be seen directly by integration by parts in (1.51). □.

Remark 9 : Higher order derivatives of k can be computed as well.

Definition 1.23 : The function J is defined on $\{v_1 \in H_1 \,;\, \|v_1\| < \varepsilon\}$ by

$$(1.56) \quad J(v_1) = I(h + k(v_1))$$

Let $p_0 \in T_{x_0}^* M$ be the unique element of $T_{x_0}^* M$ such that

$$\phi_t^h(x_0) = \pi\psi_t(x_0, p_0)$$

(p_0 exists and is unique by Theorem 1.17).

We finally have :

Theorem 1.24 : We have

(1.57) $\quad \dfrac{\partial J}{\partial v_1}(0) = 0$

Moreover if v_1', $v_1'' \in H_1$,

(1.58) $\quad \dfrac{\partial^2 J}{\partial v_1^2}(0)(v_1', v_1'') = - <p_0, B(v_1', v_1'')>$

Proof : We have

(1.59) $\quad \dfrac{\partial J}{\partial v_1}(v_1)(v_1') = <h + k(v_1), \dfrac{\partial k}{\partial v_1}(v_1)v_1'>$

(1.57) then follows from (1.50). Moreover

(1.60) $\quad \dfrac{\partial^2 J}{\partial v_1^2}(v_1)(v_1', v_1'') = <\dfrac{\partial k}{\partial v_1}(v_1)v_1'', \dfrac{\partial k}{\partial v_1}(v_1)v_1'> + <h + k(v_1), \dfrac{\partial^2 k}{\partial v_1^2}(v_1)(v_1', v_1'')>$.

Now recall that by Theorem 1.17

(1.61) $\quad h = \rho(p_0)$.

From (1.52), (1.60), we get (1.58). □.

Remark 10 : Obviously

(1.62) $\quad I(h + v_1 + k(v_1)) = I(v_1) + J(v_1)$

so that

(1.63) $\quad I''(h)(v_1', v_1'') = \langle v_1', v_1'' \rangle_{H_1} - \langle p_0, B(v_1', v_1'') \rangle$

Of course since $I'(h)$ is 0 on $T_h K_x^{x_0} = H_1$, $I''(h)$ is intrinsically defined as a bilinear form on $T_h K_x^{x_0} \times T_h K_x^{x_0}$ (this is not the case for the derivatives of higher order).

As noted by Gaveau [32] - [33], we can define extended Jacobi fields. Namely we may consider the set of the $v_1' \in H_1$ such that for any $v_1'' \in H_1$

(1.64) $\quad I''(h)(v_1', v_1'') = 0.$

It is easy to find that $v_1' \in H_1$ has to be such that for one $q \in T_{x_0}^* M$

(1.65) $\quad v_1'^j(t) = \langle p_0, [\int_0^t \sum_{i=1}^m (\phi_s^{h*-1} X_i)(x_0) v_1'^i(s) ds, (\phi_t^{h*-1} X_j)(x_0)] \rangle +$

$\langle q, (\phi_t^{h*-1} X_j)(x_0) \rangle .$

Since

$$h_s^j = <p_0, (\phi_s^{h*-1} X_j)(x_0)>$$

it is easy to check that $v_t^{'j} = \dfrac{\partial h_t^j}{\partial p} q$ is also a solution of equation (1.65), so that $v_1' = v'$. This immediately shows that

(1.66) $\quad \phi_t^{h*} \displaystyle\int_0^t (\phi_s^{h*-1} X_i)(x_0) v^i \, ds = \pi^* \dfrac{\partial \psi}{\partial p} t(p_0, x_0) q$

The variation of $\phi_t^h(x_0)$ in the direction v is then obtained as the projection of the variation of the bicharacteristic $\psi_t(x_0, p_0)$, as for classical Jacobi fields.

<u>Definition 1.25</u> : We will say that x_0 and x are not conjugate with respect to h if there is no $q \neq 0 \in T_{x_0}^* M$ such that

(1.67) $\quad \pi^* \dfrac{\partial \psi_1^*}{\partial p}(x_0, p_0) q = 0$

We now give a final result which extends the corresponding result in classical Riemannian geometry.

<u>Theorem 1.26</u> : If there is a unique $h \in K_x^{x_0}$ minimizing I on $K_x^{x_0}$, if C_1^{h,x_0} is invertible and if x_0 and x are not conjugate with respect to h, then $\bar{E}$ is a C^∞ function on a neighborhood of x.

Proof : Because of (1.67), and using the implicit function theorem, we find that a neighborhood $\mathcal{V}^1$ of p_0 in $T^*_{x_0} M$ is mapped on a neighborhood $\mathcal{V}^2$ of x in M by the mapping

(1.68) $\quad p \in \mathcal{V}^1 \to \pi\psi_t(x_0,p) \in \mathcal{V}^2$

which is a diffeomorphism of $\mathcal{V}^1$ on $\mathcal{V}^2$. By Proposition 1.16, we know that h^p defined by (1.30) exists such that

$$\pi\psi_t(x_0,p) = \phi_t^{h^p}(x_0)$$

and so for $y \in \mathcal{V}^2$, $K_y^{x_0}$ is non empty. Let $h(y)$ be an element of $K_y^{x_0}$ such that

$$I(h(y)) = \bar{E}(y).$$

Using (1.68), we find that as $x_n \to x$, $I(h(x_n))$ is a uniformly bounded sequence, so that $h(x_n)$ has a weak cluster point $\bar{h} \in K_x^{x_0}$. For n large enough, $x_n \in \mathcal{V}^2$, and so there exists $p_n \in \mathcal{V}^1$ such that $\pi\psi_1(x_0,p_n) = x_n$; moreover $p_n \to p_0$. Using (1.30), we see that $h^{p_n} \to h^{p_0} = h$, and so

(1.69) $\quad I(\bar{h}) \leq \underline{\lim}\, I(h(x_n)) \leq \overline{\lim}\, I(h(x_n)) \leq \lim I(h^{p_n}) = I(h).$

(1.69) implies that $I(\bar{h}) = I(h)$. From the uniqueness of $h \in K_x^{x_0}$ minimizing I, we find that $h = \bar{h}$ and that $I(h(x_n)) \to I(h)$ so that $h(x_n) \to h$ strongly in H. Since C_1^{h,x_0} is invertible, for n large enough, $C_1^{h(x_n),x_0}$ is also invertible. By Theorem 1.17, there is a unique $p_0(x_n) \in T^*_{x_0} M$ such that for $0 \leq t \leq 1$

(1.70) $$\phi_t^{h(x_n)}(x_0) = \pi\psi_t(x_0, p_0(x_n)).$$

$$h(x_n) = h^{p_0(x_n)}$$

The same argument as in (1.41) shows that the sequence $p_0(x_n)$ is uniformly bounded, and then has a cluster point $\bar{p}_0$. Because of (1.70), we know that

(1.71) $$\phi_t^h(x_0) = \pi\psi_t(x_0, \bar{p}_0); \quad h = h^{\bar{p}_0}$$

From Theorem 1.17, $\bar{p}_0 = p_0$. The whole sequence $p_0(x_n)$ converges to p_0. Clearly

(1.72) $$\pi\psi_1(x_0, p_0(x_n)) = x_n$$

Because of the diffeomorphism property of (1.68), we find that for n large enough,

(1.73) $$p_0(x_n) = p_n.$$

We now claim that a neighborhood $\mathcal{V}^{1,2} \subset \mathcal{V}^2$ of x exists such that if $y \in \mathcal{V}^{1,2}$, and if $p \in \mathcal{V}^1$ is uniquely determined by

(1.74) $$\pi\psi_1(x_0, p) = y$$

then

(1.75) $$\bar{E}(y) = I(h^p).$$

If this were not the case, we could find $x_n \to x$ such that $\overline{E}(x_n) < I(h^{p_n})$ where $p_n \in \mathcal{V}^1$ is defined by

(1.76) $\quad \pi\psi_1(x_0, p_n) = x_n.$

Now as we have previously seen, for n large enough, $p_n = p_0(x_n)$. By (1.70), we know that $h(x_n) = h^{p_0(x_n)}$. We then find that for n large enough

(1.77) $\quad \overline{E}(x_n) = I(h(x_n)) = I(h^{p_0(x_n)}) = I(h^{p_n}).$

A contradiction is then obtained so that (1.74), (1.75) hold. Now using the conservation of $\mathcal{H}$ under ψ, we know that

(1.78) $\quad I(h^p) = \int_0^1 \frac{|h_s^p|^2}{2} ds = \mathcal{H}(x_0, p).$

The r.h.s. is obviously a C^∞ function of p. The result immediately follows from (1.75). □.

Remark 11 : It can also be proved that a neighborhood $\mathcal{V}"$ of x exists such that if $y \in \mathcal{V}"$, $h(y)$ minimizing I on $K_y^{x_0}$ is unique, and x_0 and y are not conjugate for $h(y)$. The proof is easy by using the argument of the proof of Theorem 1.26.

II - BROWNIAN MOTION ON A RIEMANNIAN MANIFOLD AND THE CALCULUS OF VARIATIONS.

In this section, we prove some basic properties of Brownian motion with a drift on a Riemannian manifold, and develop the corresponding Malliavin calculus of variations.

A key observation was done by Malliavin [47], Eells and Elworthy [27]-[31] when they noticed that the Brownian motion x_t on a Riemannian manifold M could be lifted to the bundle of orthonormal frames N so that the lifted process u_t is itself a Markov diffusion, which is a solution of a globally defined stochastic differential equation of the type (0.23) (this is not the case for x_t since in general, M is not parallelizable).

In [48], Malliavin used this description of Brownian motion to apply the stochastic calculus of variations on u_t to obtain informations on x_t, and defined the concept of stochastic Jacobi fields. Taniguchi [65] has applied the Malliavin calculus on u_t to perform integration by parts on x_t. However, Malliavin noticed that this procedure was not fully satisfactory since extra terms appeared which were not to be expected when studying x_t. In this section, we solve this difficulty by using another invariance property of Euclidean Brownian motion, which is the invariance under rotations.

In a), the main notations are introduced. In b), the cotangent bundle T^*M is briefly described as a trivial quotient of a sub-manifold of T^*N by the orthogonal group $O(n)$ which has a Poisson action on T^*N [4].

The principal symbols of the horizontal Laplacian on N and the Laplace-Beltrami operator on M are shown to be related by this quotient procedure. In c), we show how to use the rotational invariance of the Euclidean Brownian motion, so that the Malliavin calculus of variations now gives informations on x_t.

In d), the Brownian motion x_s ($0 \leq s \leq t$) conditional on $x_t = y_0$ is defined. In e) the time reversed conditional process is also defined. The Malliavin-Eells-Elworthy's construction of these processes are shown to be related (which is a non trivial fact). In f), an expression of $\dfrac{\text{grad}_x \, p_t(x,y)}{p_t(x,y)}$ as the expectation of a stochastic integral is obtained using the results of c). In g), the conditional process x is shown to be a semi-martingale on [o,t], this by using the results of f). This answers a query of Molchanov [54]. Although an analytic proof of this fact certainly exists, it is interesting to see how the calculus of variations gives us probabilistic properties of certain processes.

a) <u>Notations and assumptions</u>.

M is now a C^∞ compact connected Riemannian manifold of dimension n.

N is the bundle of orthonormal frames in TM [42] - I (5.7). N is also compact.

x denotes the standard element of M, and u the standard element in N.

π is the canonical projection of N on M. If $u \in N$, u can be considered as a linear isometry from R^n (endowed with its canonical Euclidean structure) into $T_{\pi u}M$.

$O(n)$ denotes the group of orthogonal linear transformations of R^n, and $\mathcal{G}(n)$ its Lie algebra.

Γ is the Levi-Civita connection on N [42]-IV. Let ω be the connection form valued in $\mathcal{G}(n)$. θ denotes the canonical one-form on N valued in R^n, so that if $X \in T_u(N)$

(2.1) $\quad \theta(X) = u^{-1} \pi^* X$

From [42]- IV.2, we know that the equations of the connection Γ are given by

(2.2) $\quad d\theta = -\omega \wedge \theta$

$\quad\quad\quad d\omega = -\omega \wedge \omega + \Omega$.

In the sequel we will simplify the notations when using (2.2). Namely if $u \in N$, θ, $\theta' \in R^n$, we will write $\Omega(\theta,\theta')$, instead of $\Omega((u\theta)^*,(u\theta')^*)$ (where $(u\theta)^*$, $(u\theta')^*$ are the horizontal lifts of $u\theta, u\theta'$).

Let $Y_1,\ldots,Y_n$ be the standard horizontal vector fields on N [42]-IV. If $e_1,\ldots,e_n$ is the canonical Euclidean basis of R^n, we have :

(2.3) $\theta(Y_i) = e_i$, $\omega(Y_i) = 0$

$\mathcal{L}$ denotes the horizontal Laplacian on N, which is the second order differential operator

(2.4) $\mathcal{L} = \frac{1}{2} \sum_{1}^{m} Y_i^2$

∇ is the covariant differentiation operator on M for Γ. R denotes the curvature tensor, and S the Ricci tensor. In (2.2) Ω is the equivariant representation of R. Let J be the equivariant representation of S.

Let Δ be the Laplace-Beltrami operator on M. If $f \in C^\infty(M)$, we have :

(2.5) $\mathcal{L}(f \circ \pi) = \frac{1}{2} (\Delta f) \circ \pi$

Of course (2.5) is related to the invariance of the Laplace-Beltrami operator under the action of the structure group O(n).

b) T^*M as a quotient of T^*N.

The principal $H(x,p)$ of the operator $\frac{\Delta}{2}$ is the function defined on T^*M by :

(2.6) $H(x,p) = \frac{\|p\|^2}{2}$

(where $\|p\|$ is the Riemannian norm of p).

We now compute the principal symbol of $\mathcal{L}$ on T^*N. Namely if $u \in N$, and $\lambda \in T_u^*N$, we can write :

(2.7) $$\lambda = \sum_{i=1}^{n} a_i \theta^i + \sum_{1 \le i < j \le n} b_i^j \omega_j^i$$

The principal symbol $K(u,\lambda)$ of $\mathcal{L}$ on T^*N is then given by

(2.8) $$K(u,\lambda) = \frac{\sum_{i=1}^{n} a_i^2}{2}$$

Now the action of $O(n)$ on N extends to a symplectic action of $O(n)$ on T^*N. This action is Poisson in the sense of Arnold [4]-Appendix 5. Namely if $A \in \mathcal{G}(n)$ and if A^* is the vector field on N defined by

(2.9) $$\theta(A^*) = 0 \qquad \omega(A^*) = A$$

the action of e^{tA^*} on T^*N is associated to the Hamiltonian

(2.10) $$(u,\lambda) \in T^*N \to H_A(u,\lambda) = \langle \lambda, A^* \rangle.$$

(2.2) shows that if $\{,\}$ is the Poisson Bracket

(2.11) $$\{H_A, H_{A'}\} = H_{[A,A']}$$

so that the action of $O(n)$ is indeed Poisson.

Let N' be the submanifold of T^*N which consists of the $(u,\lambda) \in T^*N$ such that if λ is given by (2.7), then all the b_i^j are 0. N' is also given by :

$$N' = \{(u,\lambda) \; ; \; H_A(u,\lambda) = 0 \text{ for every } A \in \mathfrak{t}_j^{\cdot}(n)\}.$$

Of course N' is stable under O(n).

Now general results on Poisson actions [1], [4] show that N'/O(n) is a symplectic manifold. The reader will check easily that N'/O(n) is exactly T^*M. Let ρ be the canonical projection $N' \to T^*M$. If $(u,\lambda) \in N'$, and if λ has the representation

(2.12) $\qquad \lambda = \sum_1^n a_i \theta^i$

then $\rho(u,\lambda)$ is exactly $(\pi u, ua)$

(ua is an element of $T_{\pi u} M$ which is identified to $T^*_{\pi u} M$ by means of the metric)

Moreover the function K is obviously invariant under O(n), and if $(u,\lambda) \in N'$

(2.13) $\qquad K(u,\lambda) = H\left(\rho(u,\lambda)\right)$

From [1] - [4] we find that N' is stable under the Hamiltonian flow associated to K and that this Hamiltonian flow projects on T^*M as the Hamiltonian flow associated to H (i.e. the geodesic flow).

Conversely, if $(u,\lambda) \in N'$ and if λ is given by (2.12), consider the geodesic in M $x_t = \exp_{\pi u}(t \, u \, \lambda)$. Let u_t be the parallel translation of u along x_t. Then it can be checked that $(u_t, \sum_1^n a_i \theta^i)$ is an integral curve in N' of the Hamiltonian flow associated to K.

c) The Malliavin calculus of variations on a Riemannian manifold.

To construct the Brownian motion on a Riemannian manifold, Malliavin [47], Eells-Elworthy [27]-[31] exploited the relation (2.5). We will now use their construction.

Let Ω be the space $\mathcal{C}(R^+;R^n)$ whose standard element is ω, whose trajectory will be $w_t = (w_t^1,\ldots,w_t^n)$. Let F_t be the σ-field $\mathcal{B}(w_s|s\leq t)$, and $\{F_t\}_{t\geq 0}$ be the corresponding filtration. P denotes the Wiener measure on Ω, with $P(w_0=0)=1$. $\{F_t\}_{t\geq 0}$ will be made right continuous as in Dellacherie - Meyer [18], complete if necessary, without further mention.

dw denotes the Stratonovitch differential of w and δw its Itô differential (Meyer [52]).

On N, the topology C_K^∞ of uniform convergence of functions and their derivatives is defined as in Section 1.

b(x) denotes a C^∞ vector field on M. b'(u) is the vector field on N which is the horizontal lift of b, so that $\pi^* b' = b$, $\omega(b') = 0$. $\nabla.b$ denotes the equivariant representation of the (1,1) tensor $\nabla.b$.

For $u_0 \in N$, consider the stochastic differential equation :

(2.14) $du = b'(u)ds + Y_i(u).dw^i$

$u(0) = u_0$

Since N is compact, (2.14) has a unique solution. Moreover from (2.5), we see that $x_s = \pi u_s$ is a Markov diffusion on M, whose infinitesimal generator is $\frac{1}{2}\Delta + b$.

Malliavin [47], Eells and Elworthy [27]-[31] show that if $b = 0$, x_s is developed into the Euclidean Brownian motion $\beta_s = u_0 w_s$ (for the definition of development, see [42]-III-4).

From the results of Bismut [10]- Theorem I.4.1, Kunita [43], we now define :

<u>Definition 2.1</u> : $\phi_s(\omega,.)$ denotes the stochastic flow of C^∞ diffeomorphisms of N associated to equation (2.14).

Of course a.s , $\phi_s(\omega,.)$ depends continuously on $s \in R^+$ for the topology C_K^∞. Moreover, if $u_0 \in N$, $\phi_s(\omega,u_0)$ is the unique solution of (2.14).

We will now show how to exploit the rotational invariance of the Brownian motion w so as to transfer the results of the calculus of variations which is done on the horizontal lift u_t of x_t to results which can be directly expressed in terms of x_t.

We have the analogue of Theorem 2.1 in Bismut [12].

<u>Theorem 2.2</u> : Take $u_0 \in N$. Let u_t be the process $\phi_t(\omega,u_0)$, and $x_t = \pi u_t$.

Let $v_t = (v_t^1,\ldots,v_t^n)$ be a bounded predictable process with values in R^n. Let θ_t^v be the solution of the differential equation

(2.15) $\quad d\theta^v(t) = [(-\frac{1}{2} J + \overline{\nabla.b})\theta^v + v] ds$

$\theta^v(0) = 0$

Then if $f \in C_b^\infty(M)$, for any $t \geq 0$, the following equality holds

(2.16) $\quad E [<df(x_t), u_t\theta^v(t)>] = E [f(x_t) \int_0^t v^i \delta w^i]$

Proof : The proof starts as the proof of Theorem 2.1 in [12]. For $\ell \in R$, consider the stochastic differential equation

(2.17) $\quad du^\ell = b'(u^\ell)dt + Y_i(u^\ell)(dw^i + \ell v^i dt)$

$u^\ell(0) = u_0$

If Z_t^ℓ is the Girsanov exponential

(2.18) $\quad Z_t^\ell = \exp \{-\int_0^t \ell v^i \delta w^i - \frac{1}{2}\int_0^t (\ell v^i)^2 ds\}$

under the probability law $Z_t^\ell dP$, $w_s + \ell\int_0^s v_h \, dh$ ($0 \leq s \leq t$) is still a Brownian motion, so that

(2.19) $\quad E [Z_t^\ell f(\pi u_t^\ell)] = E [f(x_t)]$

The differential of (2.19) at $\ell=0$ is 0. Now by [12] we know that differentiation under the expectation is feasible, and that the usual rules of variation of parameters apply in (2.17). Set:

(2.20) $\quad \theta(\dfrac{du_s^\ell}{d\ell})_{\ell=0} = \bar{\theta}$

$\quad \omega(\dfrac{du^\ell}{ds})_{\ell=0} = \bar{\omega}$

From the equations of the connection (2.2), we immediately find:

(2.21) $\quad d\bar{\theta} = (\overline{\nabla b}\ \bar{\theta} - \bar{\omega}\theta(b') + v)ds + \bar{\omega}(\theta(b')ds + dw)\ ;\ \bar{\theta}(0) = 0$

$\quad d\bar{\omega} = \Omega(dw + \theta(b')ds, \bar{\theta})$

We now rewrite the first line in (2.21) in Itô's form. Clearly by [52] p. 354

(2.22) $\quad \displaystyle\int_0^t \bar{\omega}\ dw = \int_0^t \bar{\omega}\delta w + \dfrac{1}{2}\int_0^t \sum_{i=1}^n \Omega(e_i, \bar{\theta})e_i\ ds$

Now $-\sum_{i=1}^n \Omega(e_i, .)e_i$ is clearly the equivariant representation J of the Ricci tensor S, so that the first line in (2.21) writes:

(2.23) $\quad d\bar{\theta} = (-\dfrac{1}{2}J\bar{\theta} + \overline{\nabla b}\ \bar{\theta} + v\ ds) + \bar{\omega}\ \delta w$

$\quad \bar{\theta}(0) = 0$

Differentiation of (2.19) at $\ell=0$ shows that

(2.24) $\quad E\ [<df(x_t), u_t\bar{\theta}_t>] = E\ [f(x_t)\displaystyle\int_0^t v^i \delta w^i]$

A_t is now a predictable process valued in $\mathcal{G}(n)$. For $\ell \in R$, set :

(2.25) $\quad O_t^\ell = \exp \ell\, A_t$

$$w_t^\ell = \int_0^t O_s^\ell \cdot \delta w_s$$

O_t^ℓ is a predictable process and w_t^ℓ is still a Brownian motion. If u'^ℓ is the unique solution of :

(2.26) $\quad du'^\ell = b'(u'^\ell)ds + Y_i(u'^\ell) \cdot dw_t^{\ell,i}$

$$u'^\ell(0) = u_0$$

u'^ℓ has the same law as u, so that :

(2.27) $\quad E\,[f\,(\pi u'_t^\ell)] = E\,[f(x_t)].$

The differential of (2.27) at $\ell=0$ is 0. Set

(2.28) $\quad \theta(\dfrac{du'^\ell}{d\ell})_{\ell=0} = \bar{\theta}'$

$$\omega(\dfrac{du'^\ell}{d\ell})_{\ell=0} = \bar{\omega}' \;.$$

Still using (2.2), we have :

(2.29) $\quad d\bar{\theta}' = (\overline{\nabla b}\,\bar{\theta}' - \bar{\omega}'\theta(b'))ds + A\delta w + \bar{\omega}'(\theta(b')ds + dw); \bar{\theta}'(0) = 0$

$$d\bar{\omega}' = \Omega(dw + \theta(b')ds,\, \bar{\theta}')\,;\, \bar{\omega}'(0) = 0$$

The first line of (2.29) writes in Itô's form :

(2.30) $\quad d\bar{\theta}' = (-\frac{1}{2}J\bar{\theta}' + \overline{\nabla b}\,\bar{\theta}')ds + (A + \bar{\omega}')\delta w$

$\quad\quad\quad \bar{\theta}'(0) = 0$

Differentiation of (2.27) shows that

(2.31) $\quad E[<df(x_t), u_t\,\bar{\theta}'_t>] = 0$

Set :

$\quad\quad\quad \theta = \bar{\theta} + \bar{\theta}', \quad \omega = \bar{\omega} + \bar{\omega}'$

We obviously have :

(2.32) $\quad d\theta = (-\frac{1}{2}J\theta + \overline{\nabla b}\theta + v)\,ds + (A + \omega)\delta w\,;\,\theta(0) = 0$

$\quad\quad\quad d\omega = \Omega(dw + \theta(b')ds,\theta)\,;\,\omega(0) = 0.$

Moreover, by summing (2.24) and (2.31), we get

(2.33) $\quad E[<df(x_t),u_t\,\theta_t>] = E[f(x_t)\int_0^t v^i\,\delta w^i].$

Now consider the system :

(2.34) $\quad d\theta^v = [(-\frac{1}{2}J + \overline{\nabla b})\theta^v + v]ds\,;\,\theta^v(0) = 0$

$\quad\quad\quad d\omega^v = \Omega(dw + \theta(b')ds,\theta^v)\,;\,\omega^v(0) = 0$

Clearly ω_t^V is a predictable process with values in $\mathcal{Y}(n)$. Moreover (θ^V, ω^V) is exactly the solution of (2.32) with $A = -\omega^V$. From (2.33), we find (2.16).

□

Remark 1 : Of course (2.15), (2.16) can be put in a form which makes completely disappear the lift u_t of x_t. Namely observe that as pointed out in [10]-IX, [31], $u_t u_0^{-1}$ is exactly the parallel translation operator τ_t^0 along x_t. If K is a tensor field on M, we may then define the covariant differential $D_t K$ of K along x_t. Similarly if α is a vector field on M, we may define the Itô integral of α along x, by setting

$$\int_0^t \alpha \cdot \delta x = \int_0^t \tau_0^s \alpha \cdot \delta \beta$$

(recall that $\beta = u_0 w$). (2.15)-(2.16) has the following equivalent form :

Corollary : If $u_0 \in N$, set $u_t = \phi_t(\omega, u_0)$ and $x_t = \pi u_t$. Let α_t be a bounded predictable process with values in $T_{x_t} M$ (i.e for each t, $\alpha_t \in T_{x_t} M$). Let X_t be the process with values in $T_{x_t} M$ given by :

(2.35) $D X_t = (-\frac{1}{2} S X + \nabla_X b + \alpha) dt$; $X_0 = 0$

Then, if $f \in C_b^\infty(M)$, for any $t \geq 0$, the following equality holds :

(2.36) $E[<df(x_t), X_t>] = E [f(x_t) \int_0^t \alpha \delta x]$.

Remark 2 : This approach of the calculus of variations solves the question raised by Malliavin of the difficulty related to the appearance in the calculus of variations of order 1 of the first covariant derivative of the curvature tensor R (which appears in the second line of (2.21) written in Itô form). Moreover (2.36) shows that the true Jacobi fields for x_t (in the sense of Malliavin [48]) are computable by means of the Ricci tensor instead of the curvature tensor. Moreover, since the corollary of Theorem 2.2 is the analogue of Theorem 2.1 in [12] it also permits the development of the Malliavin calculus on x_t, without using the corresponding calculus on the lift u_t like in Taniguchi [65].

d) The conditional process.

We now define the process u_t conditioned on $x_t = y_d$ as a u-process (in the sense of Doob) of the initial process.

Definition 2.3 : dy denotes the Riemannian volume form on M. If $u_o \in N$, $x_t = \pi\phi_t(\omega, u_o)$, for $t > 0$, $p_t(x_o, y)dy$ is the probability law of x_t under P.

Of course, for $t > 0$, $p_t(x,y)$ in C^∞ in the variables $(x,y) \in M \times M$, and is jointly continuous in (t,x,y) for $(t > 0)$, with continuous differentials in t, x, y. This can be seen either by standard arguments on elliptic operators, or by using the Malliavin calculus.

Moreover, for any $t > 0$, $x, y \in M$, $p_t(x,y) > 0$. Clearly if $0 \leq s < t$:

(2.37) $\quad \frac{\partial}{\partial s}(p_{t-s}(x,y)) + \frac{1}{2}\Delta_x p_{t-s}(x,y) + <b(x), \frac{\partial p_{t-s}}{\partial x}(x,y)> = 0$

It follows that if $x_s = \pi\phi_s(\omega,u_0)$, for $t > 0$, and ε such that $0 \leq \varepsilon < t$, if $s \leq \varepsilon$, $p_{t-s}(x_s,y)$ is a bounded martingale. It can then be easily proved that $p_{t-s\wedge t}(x_s,y)$ is a local martingale, which is stopped at time t, and is 0 for $s \geq t$.

We then define the Brownian motion on M conditional on $x_t = y_0$. y_0 is here a fixed element of M.

Definition 2.4 : For $t > 0$, $u_0 \in N$, if $u_s = \phi_s(\omega,u_0)$, $x_s = \pi u_s$, $P^t_{u_0,y_0}$ is the unique probability measure on F_{t-} such that for any $s < t$

(2.38) $\quad \frac{dP^t_{u_0,y_0}}{dP}\bigg|_{F_s} = \frac{p_{t-s}(x_s,y_0)}{p_t(x_0,y_0)}$

Now for $s < t$, $\frac{1}{p_{t-s}(x_s,y_0)}$ is clearly a positive martingale for $P^t_{u_0,y_0}$ which has necessarily a limit as $s \uparrow\uparrow t$.

From Azencott and al [8], IX, Theorem 4.8, we know that if d is the geodesic distance in M, $a > 0$ constant C' exists such that for $s > 0$, $x',y' \in M$

$$p_s(x',y') \leq \frac{C'}{s^{n-\frac{1}{2}}} \exp -\frac{d^2(x',y')}{2s}$$

Since as $s \uparrow\uparrow t$, $\frac{1}{p_{t-s}(x_s,y_0)}$ has $P^t_{u_0,y_0}$ a.s. a limit, we find :

Proposition 2.5 : $P^t_{u_0,y_0}$ a.s, as $s \uparrow\uparrow t$, x_s converges to y_0.

We then define :

Definition 2.6 : $Q^t_{x_0,y_0}$ denotes the probability measure on $C(R^+;M)$ which is the probability law of $x_{s \wedge t}$ under $P^t_{u_0,y_0}$.

It should be pointed out that $y_0 \to Q^t_{x_0,y_0}$ is a regular conditional desintegration of the probability law of x_s ($0 \le s \le t$) conditional on $x_t = y_0$. However as in the finite dimensional case (for smooth densities), this is not an abstract desintegration of the law of x, but is expected to be a smooth desintegration. In particular results which are easy to obtain for a.e. value of y_0 will have to be proved for every value of y_0.

Of course $Q^t_{x_0,y_0}$ coïncides with the conditional Brownian motion in Molchanov [54] -5. Moreover the uniform bounds in the proof of Theorem 5.1 in [54] show easily that $Q^t_{x_0,y_0}$ depends continuously on (x_0,y_0).

Similarly $P^t_{u_0,y_0}$ can be easily shown to depend continuously on u_0,y_0.

Remark 3 : Before going into details, we need to raise a rather subtle point concerning the Malliavin calculus of variations. Consider a bounded random variable H on Ω with is C^∞ in the sense of the Malliavin calculus [12] - [46]. We know that if $x_t = \pi \phi_t(\omega,u_0)$, for $t > 0$, the law of x_t under HdP is of the form $q_t(y)dy$, where q is C^∞. Clearly

(2.39) $\quad q_t(y) = [E^{P_{u_0,y}^t} H] p_t(x_0,y) \quad$ a.e. on M.

Now if H is reasonably behaved, it can be showed that $y \to E^{P_{u_0,y}^t} H$ is a continuous function so that instead of the a.e. equality (2.37), we have equality everywhere.

We will use this fact in the sequel.

e) - <u>The time reversed conditioned process</u>.

We here define the time reversed conditional process as a u-process of a slightly modified diffusion in N. With such a definition, we absolutely need to prove that, say, parallel translation defined for the conditioned process and the time reversed process coincide in the obvious way. The proof involves non-trivial results of invariance of semi-martingales under an absolutely continuous change of probability law.

Ω' denotes another copy of Ω. The standard element of Ω' is ω'. The trajectory of ω' is $w' = (w'^1,\ldots,w'^n)$.

<u>Definition 2.7</u> : On (Ω',P), $\phi'.(\omega',.)$ is the flow of C^∞ diffeomorphisms of N associated to the stochastic differential equation :

(2.40) $\quad du' = -b'(u')dt - Y_i(u').dw'^i$

$\quad\quad\quad u'(0) = u'_0$

From the forward Fokker-Planck equation, we know that :

$$(2.41) \quad \frac{\partial p_t}{\partial t}(x_0,z) = \frac{1}{2}\Delta_z p_t(x_0,z) - \langle b(z), \frac{\partial p_t}{\partial z}(x_0,z)\rangle - (\text{div } b(z))p_t(x_0,z).$$

We now define :

<u>Definition 2.8</u> : If $u_0' \in N$ is such that $\pi u_0' = y_0$, if $y_s = \pi \phi_s'(\omega, u_0')$, $P_{u_0',x_0}^{\cdot t}$ is the probability measure on F_{t-} such that for any $s < t$:

$$(2.42) \quad \frac{dP_{u_0',x_0}^{\cdot t}}{dP}\bigg|_{F_s} = \frac{p_{t-s}(x_0,y_s)\exp-\int_0^s \text{div } b(y_h)dh}{p_t(x_0,y_0)}$$

From (2.41) we see that $P_{u_0',x_0}^{\cdot t}$ is a probability measure. The same argument as in Proposition 2.5 shows that as $s \uparrow\uparrow t$, y_s converges to x_0.

<u>Definition 2.9</u> : $Q_{y_0,x_0}^{\cdot t}$ is the probability measure on $C(R^+;M)$ which is the probability law of $y_{s\wedge t}$ under $P_{u_0',x_0}^{\cdot t}$.

Of course $P_{u_0',x_0}^{\cdot t}$ is not exactly constructed from $\phi'.(\omega',.)$ and P as P_{u_0,y_0}^t was from $\phi.(\omega,.)$ and P and so another description of $P_{u_0',x_0}^{\cdot t}$ also proves to be useful.

Namely let $q_s(y_0,z)dz$ be the probability law of $y_s = \pi\phi_s'(\omega;u_0')$ under P.

$\tilde{P}_{u_0',x_0}^{\cdot}$ is then the probability measure on F_{t-} such that for every $s < t$:

(2.43) $\quad \dfrac{dP_{u_0',x_0}^{0,t}}{dP}\bigg|F_s = \dfrac{q_{t-s}(y_s,x_0)}{q_t(y_0,x_0)}$

It can then be easily proved that

(2.44) $\quad dP_{u_0',x_0}^{'t} = \dfrac{\exp(-\int_0^t \text{div}\,b(y_s)ds)\,dP_{u_0',x_0}^{0't}}{\int \exp(-\int_0^t \text{div}\,b(y_s)ds)\,dP_{u_0',x_0}^{0't}}$

Of course $P_{u_0',x_0}^{0't}$ is the exact analogue of P_{u_0,y_0}^t and $P_{u_0',x_0}^{'t}$ is obtained from $P_{u_0',x_0}^{0't}$ by a simple change of density.

We first have an elementary result :

<u>Theorem 2.10</u> : Under Q_{x_0,y_0}^t the probability law of x_{t-s} ($s \leq t$) is the same as the law of y_s ($s \leq t$) under $Q_{y_0,x_0}^{'t}$

<u>Proof</u> : The proof is elementary by inspection of the probability law of $(x_{t_1}, x_{t_2}, \ldots, x_{t_k})$ (for $0 < t_1 < \ldots < t_k < t$). It is left to the reader.

□

When b=0, both processes x_s and y_s are constructed by developing (Malliavin [47], Eells and Elworthy [27]-[31]) one Brownian motion. We will now show how these two Brownian motions are related (this is not as obvious as the hurried reader would like to think !).

Recall that on (Ω, P)

(2.45) $\quad \beta_s = u_0 w_s$

and β_s only depend on x_h ($h \leq s$). The same property holds for

(2.46) $\quad \tau_s^o = u_s u_o^{-1}$

Since P_{u_o,y_o}^t is equivalent to P on F_s ($s < t$), we may define as well β_s and τ_s^o for $s < t$ for the measure Q_{x_o,y_o}^t.

Similarly if we set for $s < t$:

(2.47) $\quad \beta_s' = u_o' w_s'$

$\tau_s'^o = u_s' u_o'^{-1}$

β_s', $\tau_s'^o$ are unambiguously defined for Q_{y_o,x_o}^t.

From Theorem 2.10, we can construct β_s', $\tau_s'^o$ under P_{u_o,y_o}^t, for $s < t$, which are associated to $y_s = x_{t-s}$. We then have :

<u>Theorem 2.11</u> : Q_{x_o,y_o}^t a.s., for any s, s' such that $0 < s < s' < t$

(2.48) $\quad \tau_{s'}^o (\beta_{s'} - \beta_s) = \tau_{t-s'}'^o (\beta_{t-s}' - \beta_{t-s'}')$

$\tau_{s'}^o [\tau_s^o]^{-1} = \tau_{t-s'}'^o [\tau_{t-s}'^o]^{-1}$

<u>Proof</u> : Take ε,η such that $0 < \varepsilon < \eta < t$.

Let da be the Haar measure on $O(n)$. du = dx da denotes the obvious "product" measure on N.

On (Ω, P), set:

$$w_s^{\varepsilon,\eta} = w_{\eta-\varepsilon} - w_{\eta-\varepsilon-s} \quad (0 \leq s \leq \eta-\varepsilon)$$

Under P, $w_s^{\varepsilon,\eta}$ is clearly a Brownian motion. Let $\omega^{\varepsilon,\eta}$ be the element of Ω corresponding to $w^{\varepsilon,\eta}$. From Theorem 1.3.1 in [10], we know that P.a.s, for any s such that $0 \leq s \leq \eta-\varepsilon$

$$(2.49) \quad \phi_s(\omega,.) = \phi'_{\eta-\varepsilon-s}(\omega^{\varepsilon,\eta},.) \circ \phi_{\eta-\varepsilon}(\omega,.)$$

On $N \times \Omega$, we put the measure

$$(2.50) \quad dR(u,\omega) = \frac{p_\varepsilon(x_0, \pi u) p_{t-\eta}(\pi\phi_{\eta-\varepsilon}(\omega,u), y_0) du\, dP(\omega)}{p_t(x_0, y_0)}$$

Let $\{\mathcal{G}_s\}_{s \geq 0}$ be the filtration associated to the σ-fields

$$\mathcal{G}_s = \mathcal{B}(u) \vee \mathcal{B}(w_h | h \leq s)$$

For the product measure $du\, dP(\omega)$, it is trivial that w_s is a semi-martingale with respect to $\{\mathcal{G}_s\}_{s \geq 0}$. Since dR is equivalent to $du\, dP(\omega)$, we know from [18] VII that w_s is still a $\{\mathcal{G}_s\}_{s \geq 0}$ semi-martingale for dR.

Set:

$$x_s^{\varepsilon,\eta} = \pi\, \phi_s(\omega,u)$$

$$y_s^{\varepsilon,\eta} = x_{\eta-\varepsilon-s}^{\varepsilon,\eta}$$

Clearly the law of $x_s^{\varepsilon,\eta}(0 \le s \le \eta-\varepsilon)$ under R is the same as the law of $x_{s+\varepsilon}$ ($0 \le s \le \eta-\varepsilon$) under Q_{x_0,y_0}^t.

Let $\mathcal{J}(\omega,v)$ be the Jacobian at $\phi_{\eta-\varepsilon}^{-1}(\omega,v)$ of the mapping $u \to \phi_{\eta-\varepsilon}(\omega,u)$. Taking $(\omega^{\varepsilon,\eta},v)$ as new variables, and using (2.49), R writes in these new variables

$$(2.51) \quad dR'(v,\omega^{\varepsilon,\eta}) = \frac{p_\varepsilon(x_0,\pi\phi_{\eta-\varepsilon}'(\omega^{\varepsilon,\eta},v))p_{t-\eta}(\pi v,y_0)\,[\mathcal{J}(\omega,v)]^{-1}}{p_t(x_0,y_0)}\, dv\, dP(\omega^{\eta,\varepsilon})$$

Now $dR'(v,\omega^{\varepsilon,\eta})$ is equivalent to the product measure $dv\, dP(\omega^{\varepsilon,\eta})$. For this product measure, $w_s^{\varepsilon,\eta}$ is a semi-martingale with respect to $\{\mathcal{G}_s'\}_{s\ge 0}$ where

$$\mathcal{G}_s' = \mathcal{B}(v) \vee \mathcal{B}(w_h^{\varepsilon,\eta} \mid h \le s)$$

From [18]- VII, we know that $w^{\varepsilon,\eta}$ is also a semi-martingale with respect to $\{\mathcal{G}_s'\}_{s\ge 0}$ for the measure $dR'(v,\omega^{\varepsilon,\eta})$. In other words, $w^{\varepsilon,\eta}$ is a semi-martingale with respect to $\{\mathcal{G}_s''\}_{s\ge 0}$ where

$$(2.52) \quad \mathcal{G}_s'' = \mathcal{B}(\phi_{\eta-\varepsilon}(\omega,u)) \vee \mathcal{B}(w_h^{\varepsilon,\eta} \mid h \le s)$$

Since $w^{\varepsilon,\eta}$ is a semi-martingale with respect to $\{\mathcal{G}_s''\}_{s\ge 0}$, the solutions of (2.40) calculated on $w^{\varepsilon,\eta}$ are also semi-martingales with respect to $\{\mathcal{G}_s''\}_{s\ge 0}$. In particular $\phi_s'(\omega^{\varepsilon,\eta},\phi_{\eta-\varepsilon}(\omega,u))$ is a semi-martingale with respect to $\{\mathcal{G}_s''\}_{s\ge 0}$.

Set

$$\mathcal{H}'_s = \mathcal{B}(y_h^{\varepsilon,\eta} \mid h \leq s).$$

Clearly,

(2.53) $\quad y_s^{\varepsilon,\eta} = \pi\phi'_s(\omega^{\varepsilon,\eta},\phi_{\eta-\varepsilon}(\omega,u))$

From a result of Stricker [18] VII, we know that $y_s^{\varepsilon,\eta}$, which is a $\{\mathcal{G}_s^{\prime\prime\prime}\}_{s\geq 0}$-semi-martingale is also a $\{\mathcal{H}'_s\}_{s\geq 0}$-semi-martingale. Moreover, from $\phi'_s(\omega^{\varepsilon,\eta},\phi_{\eta-\varepsilon}(\omega,u))$ we get the parallel translation operators along the semi-martingale $y_s^{\varepsilon,\eta}$, which is a process adapted to $\{\mathcal{H}'_s\}_{s\geq 0}$. If

$$u_s = \phi_s(\omega,u)$$

we find from (2.49) that for $0 \leq s \leq \eta-\varepsilon$, $u_{\eta-\varepsilon-s} \circ u_{\eta-\varepsilon}^{-1}$ is exactly the $\{\mathcal{H}'_s\}_{s\geq 0}$ semi-martingale of parallel translation operators $^{\varepsilon,\eta}\tau_s^{\prime 0}$ along the $\{\mathcal{H}'_s\}_{s\geq 0}$ semi-martingale $y_s^{\varepsilon,\eta}$. Moreover (2.40) shows that :

(2.54) $\quad \displaystyle\int_0^s [^{\varepsilon,\eta}\tau_h^{\prime 0}]^{-1} \cdot dy_h^{\varepsilon,\eta} = -u_{\eta-\varepsilon} w_s^{\varepsilon,\eta} - \int_0^s [^{\varepsilon,\eta}\tau_h^{\prime 0}]^{-1} b(y_h^{\varepsilon,\eta})dh.$

If $^{\varepsilon,\eta}\tau_s^0$ is the parallel translation operator $u_s u_0^{-1}$ along the semi-martingale $x_s^{\varepsilon,\eta}$, let $\beta^{\varepsilon,\eta}$ be the process (valued in $T_{x_0^{\varepsilon,\eta}} M$)

(2.55) $\quad \beta_s^{\varepsilon,\eta} = u_0 w_s$

and let $\beta'^{\varepsilon,\eta}$ be the $\{\mathcal{H}'_s\}_{s\geq 0}$ -semi-martingale (valued in $T_{y_0^{\varepsilon,\eta}}M$)

(2.56) $\beta'^{\varepsilon,\eta}_s = u_{\eta-\varepsilon} w^{\varepsilon,\eta}_s$

We find from (2.54) that :

$$\beta'^{\varepsilon,\eta}_s = \varepsilon_{,\eta} u^0_{\eta-\varepsilon}(\beta^{\varepsilon,\eta}_{\eta-\varepsilon} - \beta^{\varepsilon,\eta}_{\eta-\varepsilon-s})$$

Since $x^{\varepsilon,\eta}_s$ ($0 \leq s \leq \eta-\varepsilon$) has the same law as $x_{s+\varepsilon}$ ($0 \leq s \leq \eta-\varepsilon$) under $Q^t_{x_0,y_0}$, the Theorem follows easily from the previous results.

□

Remark 4 : The key point in the Proof of Theorem 2.11 has been that for $\varepsilon \leq s \leq \eta$, we go back to a more standard situation where it is almost "obvious" (using time reversal on flows) that (2.48) holds.

Remark 5 : It is an obvious consequence of Liouville's theorem on the geodesic flow that $Y_1 \ldots Y_m$ preserve the measure du. It then follows that the divergence of b' with respect to du at u is exactly (div b) (πu), so that in (2.47)

(2.57) $\mathcal{J}(\omega,v) = \exp \int_0^{\varepsilon-\eta} \text{div } b(x^{\varepsilon,\eta}_s) ds$

(2.51) - (2.57) fits nicely with (2.42).

Corollary : $Q^t_{x_0,y_0}$ a.s as $s \uparrow\uparrow t$, β_s has a limit β_t, τ^0_s has a limit τ^0_t, β'_s has a limit β'_t, τ'^0_s has a limit τ'^0_t so that (2.48) still holds for $0 \leq s \leq s' \leq t$.

Proof : From (2.48) we find that for $0 < s < s' < t$

(2.58) $\quad \tau^0_s(\beta_{s'} - \beta_s) = \tau^0_s [\tau^0_{s'}]^{-1} \tau'^0_{t-s'} [\beta'_{t-s} - \beta'_{t-s'}]$

$$= \tau'^0_{t-s} [\beta'_{t-s} - \beta'_{t-s'}]$$

Now as $s' \uparrow\uparrow t$, the r.h.s. of (2.58) has a limit, so that clearly as $s' \uparrow\uparrow t$, $\beta_{s'}$ has a limit. The same proof holds for β'. Similarly the second line in (2.48) shows that as $s' \uparrow\uparrow t$, $\tau^0_{s'}$, $\tau'^0_{s'}$ have limits.

$\square$

f) - **An expression for** $\dfrac{\text{grad}_{x_0} \, p_t(x_0,y_0)}{p_t(x_0,y_0)}$

We now will express $\dfrac{\text{grad}_{x_0} \, p_t(x_0,y_0)}{p_t(x_0,y_0)}$ as the expectation of a stochastic integral.

We first start with a crude estimate on β for $Q^t_{x_0,y_0}$.

Proposition 2.12 : There are constants $C, A > 0$ such that for any $x_0, y_0 \in M$, $t > 0$, $a > 0$

(2.59) $\quad Q^t_{x_0,y_0} [\sup_{0 \leq s \leq t} |\beta_s| \geq a] \leq C e^{\frac{A}{t} - \frac{a^2}{4t}}$

Proof : Clearly if $u_o \in N$ is such that $\pi u_o = x_o$, if $x_s = \pi \phi_s(\omega, u_o)$

$$(2.60) \quad Q^t_{x_o,y_o}[\sup_{0 \le s \le t/2} |\beta_s| \ge \tfrac{a}{2}] = \frac{E^P[1_{\sup_{0 \le s \le t/2}|w_s| \ge \tfrac{a}{2}} p_{\tfrac{t}{2}}(x_{\tfrac{t}{2}}, y_o)]}{p_t(x_o, y_o)}$$

Now, from the estimate in [8] IX, Theorem IV.8, for $\chi > 0$

$$(2.61) \quad p_{t/2}(x_t, y_o) \le \frac{C}{t^{n-1/2}}$$

$$p_t(x_o, y_o) \ge e^{\frac{-d^2(x_o, y_o) - \chi}{2t}}$$

Since :

$$(2.62) \quad P[\sup_{0 \le s \le t/2} |w_s| \ge \tfrac{a}{2}] \le 2\, e^{-\frac{a^2}{4t}}$$

we see from (2.61) that :

$$(2.63) \quad Q^t_{x_o,y_o}[\sup_{0 \le s \le t/2} |\beta_s| \ge \tfrac{a}{2}] \le C\, e^{\frac{A}{t} - \frac{a^2}{4t}}$$

Since div b is bounded, a similar argument will prove that :

$$(2.64) \quad Q^t_{x_o,y_o}[\sup_{0 \le s \le t/2} |\beta'_s - \beta'_{t/2}| \ge \tfrac{a}{2}] \le C\, e^{\frac{A}{t} - \frac{a^2}{4t}}$$

From Theorem 2.11 and its Corollary, we know that for $s \ge t/2$

$$(2.65) \quad |\beta_s - \beta_{t/2}| = |\beta'_{t-s} - \beta'_{t/2}|$$

From (2.63)- (2.65), the result follows.

□

Definition 2.13 : Take $u_0 \in N$. On (Ω, P), set $u_s = \phi_s(\omega, u_0)$ and $x_s = \pi u_s$. E_s is the predictable process of linear mappings of $T_{x_0} M$ in $T_{x_s} M$ defined by

(2.66) $\quad D E_s = (-\frac{1}{2} SE + \nabla_E b) ds$

$\quad E(0) = I$

E'_s is the process of linear mappings of $T_{x_0} M$ into $T_{x_0} M$ given by :

(2.67) $\quad E'_s = [\tau_s^0]^{-1} E_s$

Clearly,

(2.68) $\quad d E'_s = [\tau_s^0]^{-1} [-\frac{1}{2} S \tau_s^0 E'_s + \nabla_{\tau_s^0 E'_s} b_s] ds$

$\quad E'_s(0) = I$

From the corollary of Theorem 2.11, we see that for $Q^t_{x_0, y_0}$, E_s and E'_s are unambiguously defined on $[0, t]$. Since S and b are bounded, E_s and E'_s are uniformly bounded for $s \leq t$ (independently of ω).

Consider now the process

(2.69) $\quad \int_0^s \tilde{E}'_h \, \delta \beta_h$

On (Ω, P), β is a Brownian martingale so that (2.69) is well defined.

However, for $Q^t_{x_0,y_0}$, we know that β_h is a semi-martingale for $h < t$ (since $Q^t_{x_0,y_0}$ is equivalent to the law of $x_{\cdot}$ under P on F_s for $s < t$) but we still do not know that the semi-martingale property holds on $[0,t]$. However on (Ω, P) we can write :

$$(2.70) \quad \int_0^s \tilde{E}'_h \, \delta\beta_n = \tilde{E}'_s \, \beta_s - \int_0^s \frac{d\tilde{E}'_h}{dh} \beta_h \, dh.$$

Now the r.h.s. of (2.70) is unambiguously defined on $[0,t]$ for $Q^t_{x_0,y_0}$. Still, by a provisional abuse of notation, we will use the notation of the l.h.s. of (2.70) on $Q^t_{x_0,y_0}$.

We now have the key result :

Theorem 2.14 : For every $x_0, y_0 \in M$, $t > 0$,

$$(2.71) \quad \frac{\operatorname{grad}_{x_0} p_t(x_0,y_0)}{p_t(x_0,y_0)} = \frac{E^{Q^t_{x_0,y_0}} \int_0^t \tilde{E}'_s \, \delta\beta_s}{t}$$

Proof : We first prove that for each x_0, t, (2.71) holds a.e. in y_0. Take $u_0 \in N$ such that $\pi u_0 = x_0$. Take $X \in T_{x_0} M$, and let X' be its horizontal lift at u_0. Let $\ell \to u^\ell$ be a C^∞ curve such that $u^0 = u_0$, $\frac{du^\ell}{d\ell}\big|_{\ell=0} = X'$. To compute $\frac{d}{d\ell} \phi_t(\omega, u^\ell)\big|_{\ell=0}$, we still use the equation of the connection(2.2) so that if :

$$(2.72) \quad \theta \left[\frac{d}{d\ell} \phi_s(\omega, u^\ell) \right]_{\ell=0} = \theta"$$

$$\omega\left(\frac{d}{d\ell}\phi_s(\omega, u^\ell)\right)_{\ell=0} = \omega"$$

then

(2.73) $d\theta" = \overline{\nabla b}\,\theta" + \omega"dw$; $\theta"(o) = u_o^{-1} X$

$d\omega" = \Omega(dw + \theta(b')ds, \theta")$; $\omega"(o) = 0$

By introducing orthogonal variations of w as in the proof of Theorem 2.2, we easily find that for $t > 0$, $f \in C^\infty(M)$,

(2.74) $<\frac{\partial}{\partial x}\int p_t(x_0,y)f(y)dy, X> = E<\frac{\partial f}{\partial x}(x_t), E_t X>$

Now, if Y_s is the process

(2.75) $Y_s = \frac{s}{t} E_s X$

it is obvious that

(2.76) $DY_s = [-\frac{1}{2} S Y + \nabla_Y b + \frac{E_s X}{t}] ds$

$Y_o = 0$

and moreover, $Y_t = E_t X$. From the Corollary of Theorem 2.2, we get

(2.77) $E <\frac{\partial f}{\partial x}(x_t), E_t X> = E\,[f(x_t) \int_0^t \frac{<E_s X, \delta x>}{t}]$

so that

(2.78) $\quad \langle \frac{\partial}{\partial x} \int p_t(x_0, y) f(y) dy, X \rangle = E [f(x_t) \int_0^t \frac{\langle \tilde{E}' \delta \beta, X \rangle}{t}]$

Now $\int_0^t \langle \tilde{E}' \delta \beta, X \rangle$ is integrable for P, so that for a.e. y_0, $\int_0^t \langle \tilde{E}' \delta \beta, X \rangle$ is integrable for Q_{x_0, y_0}^t, and so:

(2.79) $\quad E [f(x_t) \int_0^t \frac{\langle \tilde{E}' \delta \beta, X \rangle}{t}] = \int p_t(x_0, y) \frac{E^{Q_{x_0,y}^t} \int_0^t \langle \tilde{E}' \delta \beta, X \rangle}{t} f(y) dy$

(and of course the integral in the r.h.s. of (2.79) makes sense). From (2.78), (2.79), we find that:

(2.80) $\quad \frac{\text{grad}_x p_t(x_0, y_0)}{p_t(x_0, y_0)} = \frac{E^{Q_{x_0,y_0}^t} \int_0^t \tilde{E}' \delta \beta}{t} \quad y_0 \text{ a.e}$

The l.h.s of (2.80) being continuous in y_0, to prove (2.71), we only need to prove the continuity of the r.h.s. of (2.80).

We claim that as $s \uparrow\uparrow t$

(2.81) $\quad E^{Q_{x_0,y_0}^t} \int_0^s \tilde{E}' \delta \beta \to E^{Q_{x_0,y_0}^t} \int_0^t \tilde{E}' \delta \beta$

uniformly in $y_0 \in M$. Namely from (2.70), we know that:

(2.82) $\quad |\int_s^t \tilde{E}' \delta \beta| \leq C|t-s| \sup_{0 \leq h \leq t} |\beta_h| + C |\beta_t - \beta_s|$

From Proposition 2.12, $E^{Q_{x_0,y_0}^t} \sup_{0 \leq s \leq t} |\beta_s|$ is uniformly bounded (as $y_0 \in M$). Moreover, since from the Corollary of Theorem 2.11, $|\beta_t - \beta_s| = |\beta'_{t-s}|$, we only need to prove that

$E^{Q_{x_0,y_0}^t} |\beta'_h| \to 0$ uniformly as $h \to 0$. Using time reversal, this is basically equivalent to proving that $E^{Q_{x_0,y_0}^t} |\beta_h| \to 0$ as $h \to 0$. Now for $h \leq t/2$, $p_{t-h}(x',y_0)$ is bounded so that

$$E^{Q_{x_0,y_0}^t} |\beta_h| \leq C \sqrt{h}.$$

To prove the continuity in y_0 of the r.h.s. of (2.80), we only need to prove that for $s < t$ the l.h.s. of (2.81) is continuous in y_0. Now, clearly

(2.83) $$E^{Q_{x_0,y_0}^t} \int_0^s \tilde{E}'\delta\beta = \frac{E^P \int_0^s \tilde{E}'\delta\beta \; p_{t-s}(x_s,y_0)}{p_t(x_0,y_0)}$$

The continuity of (2.83) in y_0 is then trivial.

□

Remark 6 : Formula (2.71) is the natural extension of the corresponding formula for the Euclidean Brownian motion (for which $\int_0^t \tilde{E}'_s \delta\beta_s = y_0 - x_0$!). Moreover (2.71) is no mystery. Namely if $\square = d\delta + \delta d$ is the Riemann-Kodaira operator, if we set

(2.84) $$\text{grad}_x p_s(x,y) = q_s(x,y)$$

by differentiating (2.37), we get

$$(2.85) \quad \frac{\partial}{\partial s}(q_{t-s}(x,y)) - \frac{1}{2}\Box_x q_{t-s}(x,y) + <\nabla_. b(x), q_{t-s}(x,y)> +$$

$$+ <b, \nabla_. q_{t-s}(x,y)> = 0$$

From Weitzenböck's formula

$$(2.86) \quad \Box_x q_{t-s}(x,y) = -\nabla_{\pi^*Y_i}^* \nabla_{\pi Y_i}^* q + Sq$$

From (2.85), (2.86), it follows easily that $\tilde{E}_s q_{t-s}(x_0, y_0)$ is a martingale on (Ω, P), so that for $s < t$, $\dfrac{\tilde{E}_s q_{t-s}(x_s, y_0)}{p_{t-s}(x_s, y_0)}$ is a martingale for Q_{x_0, y_0}^t.

Now (2.38) and a trivial use of Girsanov's transformation show that if $\bar{\beta}_s$ is given by

$$(2.87) \quad \bar{\beta}_s = \beta_s - \int_0^s \frac{\tau_h^{0^{-1}} q_{t-h}(x_h, y_0)}{p_{t-h}(x_h, y_0)} dh$$

then under Q_{x_0, y_0}^t, $\bar{\beta}_s$ is a Brownian martingale.

From these two facts, it follows that if $s < t$

$$(2.88) \quad E^{Q_{x_0, y_0}^t} \int_0^s \tilde{E}'\delta\beta = \frac{s\ \text{grad}_x p_t(x_0, y_0)}{p_t(x_0, y_0)}$$

Remark 7 : As pointed out by Molchanov [54], there is no guarantee that under $Q^t_{x_0,y_0}$, x_s is a semi-martingale on $[0,t]$. Another form of this statement is that (2.87) makes sense only for $s < t$.

g) - Semi-martingale property of the conditional process on $[0,t]$.

Using the results of the previous sections, we now prove that for $Q^t_{x_0,y_0}$ all the processes which have been considered are semi-martingales on $[0,t]$.

Theorem 2.15 : For any $x_0, y_0 \in M$, $t > 0$, for $Q^t_{x_0,y_0}$, $\beta_s (0 \le s \le t)$ is a semi-martingale on $[0,t]$, whose Itô decomposition on $[0,t]$ writes

$$(2.89) \quad \beta_s = \int_0^s \frac{\tau_0^h \, \text{grad}_x \, p_{t-h}(x_h, y_0)}{p_{t-h}(x_h, y_0)} \, dh + \bar{\beta}_s$$

where $\bar{\beta}_s$ $(0 \le s \le 1)$ is a Brownian martingale for $Q^t_{x_0,y_0}$.

Proof : We only need to prove that

$$(2.90) \quad E^{Q^t_{x_0,y_0}} \int_0^t \frac{|\text{grad}_x p_{t-h}(x_h, y_0)|}{p_{t-h}(x_h, y_0)} \, dh < +\infty$$

Indeed, we have seen that (2.89) is valid for $s < t$. If we prove (2.90), we will have shown that $\beta_s - \bar{\beta}_s$ has a bounded variation on $[0,t]$, and the proof will be finished.

Now from Theorem 2.14 and (2.70), we know that

(2.91) $$\frac{|\text{grad } p_{t-h}(x',y_0)|}{p_{t-h}(x',y_0)} \leq \frac{1}{t-h} E^{Q_{x',y_0}^{t-h}}[\sup_{0\leq s\leq t-h} |\beta_s'|]$$

Clearly, we only need to prove the finiteness of

$$E^{Q_{x_0,y_0}^{t}} \int_{t/2}^{t} \frac{|\text{grad } p_{t-h}(x_h,y_0)|}{p_{t-h}(x_h,y_0)} \, dh$$

Using (2.91), this expression is obviously dominated by

(2.92) $$\frac{\int_M [\int_{t/2}^{t} p_h(x_0,z) \frac{E^{Q_{z,y_0}^{t-h}}[\sup_{0\leq s\leq t-h} |\beta_s'|]}{t-h} p_{t-h}(z,y_0) dh] \, dz}{p_t(x_0,y_0)} =$$

$$E^{Q_{x_0,y_0}^{t}} \int_{t/2}^{t} \frac{[\sup_{0\leq s\leq t-h} |\beta_s'|] \, dh}{t-h}$$

From Theorem 2.10, the r.h.s. of (2.92) is exactly

(2.93) $$E^{Q_{y_0,x_0}^{t}} \int_0^{t/2} \frac{\sup_{0\leq s\leq h} |\beta_s'| \, dh}{h} = E^P [\int_0^{t/2} \frac{\sup_{0\leq s\leq h} |\beta_s'|}{h} \frac{p_{t-h}(x_0,y_s)}{p_t(x_0,y_0)} \exp(-$$

$$-\int_0^t \text{div } b(y_s)ds) \, dh]$$

For $0 \leq h \leq t/2$, we may bound the last terms in the r.h.s. of (2.93). Moreover :

$$(2.94) \quad E^P \int_0^{t/2} \frac{\sup_{0\leq s\leq h} |\beta'_s| \, ds}{h} \leq C \int_0^{t/2} \frac{\sqrt{h} \, dh}{h} < +\infty.$$

The proof is finished.

□

Remark 8 : Using (2.44), and the invariance of the semi-martingale property under an absolutely continuous change of measure [18], VII, a similar result holds on β'_t. Moreover, for $P^t_{u_0,y_0}$, since $w_h = u_0 \beta_h$, w_h is also a semi-martingale. The flow $\phi.(\omega,.)$ can then be explicitly constructed for $P^t_{u_0,y_0}$ on $[0,t]$. It suffices to consider again the equation (2.14) for $P^t_{u_0,y_0}$ where w is no longer a Brownian motion but a more general continuous semi-martingale, and use the results of Kunita [43]. Of course for $P^t_{u_0,y_0}$,

$$(2.95) \quad x_s = \pi \, \phi_s(\omega, u_0) \qquad 0 \leq s \leq t$$

Moreover, if $u_t = \phi_t(\omega, u_0)$, then u_t is the unique solution of the stochastic differential equation :

$$(2.96) \quad du = \{b'(u) + [\frac{\text{grad}_x p_{t-s}(\pi u, y_0)}{p_{t-s}(\pi u, y_0)}]'\} \, ds + Y_i(u) . u_0^{-1} \, d\bar{\beta}^i$$

$$u(0) = u_0$$

where $[\frac{\text{grad}_x p_{t-s}}{p_{t-s}}]'$ is the horizontal lift of $\frac{\text{grad}_x p_{t-s}}{p_{t-s}}$. Indeed, (2.96) is trivially true for $s < t$. Since when integrated both sides have limits as $s \uparrow\uparrow t$, the result holds on $[0,t]$.

In particular under $Q^t_{x_0,y_0}$, x_s ($0 \le s \le t$) is a semi-martingale in the sense of Schwartz [73].

The same sort of results holds for $P'^t_{u'_0,x_0}$, $Q'^t_{y_0,x_0}$. In particular under $P'^t_{u'_0,x_0}$ the flow ϕ' can be constructed on $[0,t]$.

We will freely use these facts without further mention.

III - CONDITIONAL DIFFUSIONS AND CONDITIONAL FLOWS : THE BASIC ESTIMATES IN THE ELLIPTIC CASE.

As pointed out in the introduction, before we can use a local parametrization of the set of paths connecting x_0 and y_0, we need to establish some global large deviations results on the corresponding conditional process.

It is at this stage that we use the ellipticity of the generator of the diffusion, since in this case, the global estimates of Varadhan [69], [70], Molchanov [54], Azencott [6], [8] give us lower and upper bounds for $p_t(x_0, y_0)$ of the type

$$(3.1) \qquad e^{-\frac{d^2(x_0, y_0) - \chi}{2t}} \leq p_t(x_0, y_0) \leq \frac{e^{-\frac{d^2(x_0, y_0)}{2t}}}{t^N}$$

The main obstacle to the extension of the techniques of Section 4 to more general hypoelliptic diffusions which verify the assumption H2 of Section 1 is that, to our knowledge, no uniform local estimate of the type (3.1) exists for the moment for these diffusions.

To establish the large deviation results on the conditional process on the time interval [0,t] (which is later rescaled to [0,1]) we first establish the corresponding property on $[0,\frac{t}{2}]$, this by using (3.1), Varadhan's theory [69], [70] and Azencott's results [6] on the unconditioned process. The estimates on $[t/2,t]$ are then obtained using time reversal.

The situation is slightly complicated by the fact that we use the description of Malliavin [47], Eells-Elworthy [27],[31] of elliptic diffusions, this in order to obtain geometrically meaningful quantities in Section 4. This makes the time reversal arguments more involved.

In a), we show how to reduce the problem to the compact case as in Molchanov [54], using the precise estimates of Azencott [8]. In b) the basic estimates on the behavior of $\dfrac{\mathrm{grad}_x\, p_t(x_0,y_0)}{p_t(x_0,y_0)}$ as $t \downarrow 0$ are proved. In c), a large deviation result is proved on a stochastic flow.

a) Reduction to the compact case

M is now a C^∞ connected Riemannian manifold, which we assume to be complete. Δ, b, $p_t(x,y)$, $\bar{P}^t_{u_0,y_0}$ are otherwise as in Section 2.

We will show how to reduce the search of the Taylor's series for $p_t(x_0,y_0)$ to the case where M is compact, this by using a precise estimate of $p_t(x_0,y_0)$ due to Azencott [8].

x_0, y_0 are two fixed elements of M. We do the basic assumption that there is one single minimizing geodesic γ_s which connects x_0 and y_0 so that

(3.2) $\gamma(0) = x_0, \quad \gamma(1) = y_0$

and that x_0 and y_0 are not conjugate along γ [41] - p.115.

Definition 3.1.

For $\varepsilon > 0$, $\mathcal{U}^\varepsilon$ denotes the open set in M defined by

$$\mathcal{U}^\varepsilon : \{z \in M \; ; \; d(z,\gamma) < \varepsilon\}$$

$\mathcal{U}^\varepsilon$ is then a tubular neighborhood of γ in M.

We have :

Proposition 3.2.

There exists $\chi > 0$ such that as $t \downdownarrows 0$

(3.3) $Q^t_{x_0, y_0}$ (for one $s \in [0,t]$, $x_s \notin \mathcal{U}^\varepsilon$) $\leq e^{-\frac{\chi}{t}}$

Proof. Clearly if Q_{x_0} denotes the (non conditioned) law of the process x_s starting at x_0, and if T^ε is the first time where x_s enters $c_{\mathcal{V}^\varepsilon}$, we have

$$(3.4) \qquad Q^t_{x_0, y_0}(T^\varepsilon \leq \tfrac{t}{2}) = \frac{E^{Q_{x_0}}(T^\varepsilon \leq t/2 \,;\, p_{t/2}(x_{t/2}, y_0))}{p_t(x_0, y_0)}$$

(if explosion has taken place before $t/2$, $p_{t/2}(x_{t/2}, y_0)$ is assumed to be 0). From Varadhan's estimate [8], [69], [70], we know that for any $\eta > 0$

$$(3.5) \qquad p_t(x_0, y_0) \geq e^{-\frac{d^2(x_0, y_0) + \eta}{2t}}$$

From Azencott's estimate [8] VIII Proposition 4.4., since M is complete we know that for $r \geq d(x_0, y_0)$, there is $N > 0$ such that for $x' \in M$

$$(3.6) \qquad p_{t/2}(x', y_0) \leq \frac{C}{t^N} \exp - \frac{d^2(x', y_0) \wedge r^2}{t}$$

From (3.4)-(3.6), we get

$$(3.7) \qquad Q^t_{x_0, y_0}(T^\varepsilon \leq t/2) \leq \frac{C}{t^N} \exp\{\frac{d^2(x_0, y_0) + \eta}{2t}\} \; E^{Q_{x_0}}[T^\varepsilon \leq t/2 \,;\, \exp - \frac{d^2(x_{t/2}, y_0) \wedge r^2}{t}]$$

Now from the results of Varadhan [69]-[70] (see also Azencott [6]), we find that since $(T^\varepsilon \leq t/2)$ is closed in $C(R^+; M)$

(3.8) $\varlimsup t \log E^{Q_{x_0}}[T^\varepsilon \leq t/2 \,;\, \exp - \frac{d^2(x_{t/2}, y_0) \wedge r^2}{t}]$

$$\leq - \inf_{x \in C_\varepsilon} [\frac{\int_0^{1/2} |\dot{x}|^2 ds}{2} + d^2(x_{t/2}, y_0) \wedge r^2]$$

where C_ε is the set of continuous mappings $t \in [0,1/2] \to x_t$ such that $x(0) = x_0$, whose derivative $\dot{x}$ is square integrable, and moreover for one t such that $0 \leq t \leq 1/2$, $x_t \in C\mathcal{U}^\varepsilon$. Now if $x \in C_\varepsilon$

(3.9) $\int_0^{1/2} \frac{|\dot{x}|^2}{2} ds + d^2(x_{1/2}, y_0) \wedge r^2$

$$\geq [\int_0^{1/2} \frac{|\dot{x}|^2}{2} ds + d^2(x_{1/2}, y_0)] \wedge r^2$$

It is easy to check that since γ is the unique minimizing geodesic connecting x_0 and y_0, there is $\delta > 0$ such that for $x \in C_\varepsilon$

(3.10) $\int_0^{1/2} \frac{|\dot{x}|^2}{2} ds + d^2(x_{1/2}, y_0) \geq \frac{d^2(x_0, y_0)}{2} + \delta$

Since $r \geq d(x_0, y_0)$, we get from (3.9), (3.10) that for $t > 0$

(3.11) $E^{Q_{x_0}}[T^\varepsilon \leq t/2 \,;\, \exp - \frac{d^2(x_{t/2}, y_0) \wedge r^2}{t}] \leq \exp\{-\frac{d^2(x_0, y_0)}{2t} - \frac{\delta}{2t}\}$

By taking $\eta = \delta/4$, we see from (3.7), (3.11) that (3.7) is dominated by $e^{-\frac{\chi}{t}}$ with $\chi < \frac{\delta}{4}$.

Similarly, we can dominate $Q'^t_{y_0, x_0}[T^\varepsilon \leq t/2]$ by $e^{-\chi/t}$. From Theorem 2.10, we can dominate $Q^t_{x_0, y_0}[t/2 \leq T^\varepsilon \leq t]$ by $e^{-\chi/t}$. Proposition 3.2 is now obvious. □

Since what we want to estimate for $Q^t_{x_0,y_0}$ should be a Taylor expansion in the variable t, (3.3) can clearly be neglected.

We claim that it is possible to embed isometrically $\mathcal{V}^{\varepsilon/4}$ in a connected C^∞ compact Riemannian manifold (say a sphere) so that γ is still the unique minimizing geodesic from x_0 to y_0.

To do this, we inflate the metric on $\mathcal{V}/\mathcal{V}^{\varepsilon/4}$ so that any curve connecting x in $\mathcal{V}^{\varepsilon/4}$ and y in $^C\mathcal{V}^{\varepsilon/2}$ has length $> d(x_0,y_0)$.

Now $\mathcal{V}^\varepsilon$ can be embedded as an open set in a connected compact manifold M'. The metric on $\mathcal{V}^{\varepsilon/2}$ can be smoothly extended to M. It is then trivial to check that γ is still the unique minimizing geodesic connecting x_0 and y_0 in M'.

Of course x_0 and y_0 are still non conjugate along γ.

b) - <u>The basic estimates.</u>

From now on, we assume that M is compact. We adopt all the notations used in Section 2.

$\mathcal{V}^\varepsilon$ has been defined in Definition 3.1.

Let $C(x_0)$ and $C(y_0)$ be the cut loci of x_0 and y_0 [41], p. 128. Since $C(x_0)$ and $C(y_0)$ are closed sets, and since $\gamma \cap (C(x_0) \cup C(y_0)) = \emptyset$, we may and we will assume that $\varepsilon > 0$ has been chosen to be small enough so that

(3.12) $\mathcal{V}^{2\varepsilon} \cap (C(x_0) \cup C(y_0)) = \emptyset$

In particular any $x \in \mathcal{V}^{2\varepsilon}$ can be connected to x_0 or y_0 by a single minimizing geodesic (at speed equal to 1).

As in Azencott [6]-[8], we first rescale time so that everything takes place on the time interval [0,1]. Moreover, for notational convenience, we will express the various random variables as functions of dw (instead of writing them as functions of w or ω).

<u>Definition 3.3</u> : For $t \in R^+$, on (Ω,P), $\psi^t.(\sqrt{t}\, dw,.)$ denotes the flow of diffeomorphisms of N associated to the stochastic differential equation

(3.13) $du = tb'(u)ds + \sqrt{t}\, Y_i(u).dw^i$

$u(o) = \bar{u}$

Using the results of [10]. I, we may and we will assume that $\psi_s^{\varepsilon^2}(\varepsilon dw, \bar{u})$ is also a smooth function of $(\varepsilon, s, \bar{u})$.

Clearly, for one given $t > 0$, $\psi^t.(\sqrt{t}\, dw,.)$ has the same law as $\phi_t.(\omega,.)$ under P.

<u>Definition 3.4</u> : For $t > 0$, $u_0 \in N$, if $u_s^t = \psi_s^t(\sqrt{t}\, dw, u_0)$, $x_s^t = \pi u_s^t$, $\bar{P}_{u_0,y_0}^t$ is the unique probability measure on F_1^- such that for any $s < 1$

(3.14) $$\left.\frac{d\bar{p}^t_{u_0,y_0}}{dP}\right|_{F_s} = \frac{p_{t(1-s)}(x^t_s,y_0)}{p_t(\pi u_0,y_0)}$$

If $\bar{x}_0 = \pi u_0$, $\bar{Q}^t_{\bar{x}_0,y_0}$ is the law of x^t_s ($0 \leq s < 1$) under $\bar{P}^t_{u_0,y_0}$.

Instead of (3.13), we may consider the stochastic differential equation

(3.13') $\quad du' = -tb'(u')ds - \sqrt{t}\, Y_i(u') . dw'^i$

$\qquad u'(0) = u'_0$

and the associated flow $\psi'^t . (\sqrt{t}\, dw', .)$

The probability measure $\bar{P}'^t_{u'_0,x_0}$, $\bar{Q}'^t_{y_0,x_0}$ are defined as in Definition 2.8 and 2.9 with the obvious changes.

We will use the notation u^t_s, x^t_s as in Definition 3.4. Similarly, we set

(3.15) $\quad u'^t_s = \psi^t_s(\sqrt{t}\, dw, u'_0)$

$\qquad y'^t_s = \pi u'^t_s .$

<u>Definition 3.5</u> : β_s, β'_s are the processes with values in $T_{\pi u_0} M$ and $T_{\pi u'_0} M$ given by

(3.16) $\quad \beta_s = u_0 w_s \;;\; \beta'_s = u'_0 w'_s \qquad 0 \leq s \leq 1.$

From the results of Section 2, we know that w, w', β, β' are semi-martingales for all the considered probability measures.

__Definition 3.6__ : For $\bar{x}_0 \in \mathcal{V}^{2\varepsilon}$, $n(\bar{x}_0)$ denotes the unique vector in $T_{\bar{x}_0} M$ such that $\exp_{\bar{x}_0} sn(\bar{x}_0)$ (0≤s≤1) is the unique minimizing geodesic connecting $\bar{x}_0$ and y_0.

$n(\bar{x}_0)$ is easily seen to be a C^∞ function of $\bar{x}_0$ on $\mathcal{V}^\varepsilon$.

The key result for establishing a large deviation property on the flow $\psi_\cdot^t(\sqrt{t}\, dw,.)$ is as follows :

__Theorem 3.7__ : For any ε' > 0, there is χ' > 0 such that for any $\bar{x}_0 \in \mathcal{V}^\varepsilon$

(3.17) $\quad \bar{Q}_{\bar{x}_0,y_0}^t (\sup_{0 \leq s \leq 1} |\sqrt{t}\, \beta_s - sn(\bar{x}_0)| \geq \varepsilon') \leq e^{-\frac{\chi'}{t}}$

As t ↓↓ 0, the functions $\bar{x}_0 \to E^{\bar{Q}_{\bar{x}_0,y_0}^t}(\sqrt{t}\, \beta_1)$

converge uniformly on $\mathcal{V}^\varepsilon$ to the function $\bar{x}_0 \to n(\bar{x}_0)$

__Proof__ : From Proposition 2.12, we know that :

(3.18) $\quad \bar{Q}_{\bar{x}_0,y_0}^t (\sup_{0 \leq s \leq 1} |\sqrt{t}\, \beta_s | \geq a) \leq Ce^{\frac{A}{t} - \frac{a^2}{4t}}$

The second part of the Theorem will then obviously follow from (3.17), (3.18) and Cauchy-Schwarz's inequality.

To prove (3.17), we first estimate

(3.19) $\quad \bar{Q}^{t}_{\bar{x}_0, y_0}(\sup_{0 \leq s \leq 1/2} | \sqrt{t}\, \beta_s - sn(\bar{x}_0) | \geq \frac{\varepsilon'}{2})$

For $f \in L^2(R^+; R^n)$, $u'_0 \in N$ such that $\pi u'_0 = \bar{x}_0$, consider the differential equation

(3.20) $\quad du^f = \sum_{1}^{n} Y_i(u^f) f^i\, ds$

$\quad u^f(0) = u'_0$

If

(3.21) $\quad C^f_s = \int_0^s u'_0 f_h\, dh$

C^f_s is the development in $T_{\bar{x}_0} M$ of $\bar{x}^f_s = \pi u^f_s$ in the sense of [42] -II. We now claim that $\delta > 0$ exists such that for any $\bar{x}_0 \in \mathcal{V}^\varepsilon$, and $f \in L^2([0,1/2]; R^n)$ such that

(3.22) $\quad \sup_{0 \leq s \leq 1/2} |C^f_s - sn(\bar{x}_0)| \geq \varepsilon'/4$

then

(3.23) $\quad \int_0^{1/2} \frac{|f|^2}{2}\, ds + d^2(\bar{x}^f_{1/2}, y_0) \geq \frac{d^2(\bar{x}_0, y_0)}{2} + \delta$.

If this were not the case, we could find $\bar{x}^n_0 \in \mathcal{V}^\varepsilon$, $f^n \in L^2([0,1/2]; R^n)$ such that (3.22) would hold with respect to f^n and moreover

(3.24) $$\int_0^{1/2} \frac{|f^n|^2 ds}{2} + d^2(\bar{x}^n_{1/2}, y_0) - d^2(\bar{x}^n_0, y_0) \to 0$$

Now $\bar{x}^n_0$ has a cluster point $\bar{x}_0$ in $\mathcal{V}^{2\varepsilon}$. f_n remaining bounded in $L_2([0,1/2];R^n)$ has a weak cluster point f. From (3.22) we find :

(3.25) $$\sup_{0 \leq s \leq 1/2} |C^f_s - sn(\bar{x}_0)| \geq \varepsilon'/4$$

From Theorem 1.1 and (3.24), we get :

(3.26) $$\int_0^{1/2} \frac{|f|^2 ds}{2} + d^2(x^f_{1/2}, y_0) - \frac{d^2(\bar{x}_0, y_0)}{2} \leq 0$$

so that equality necessarily holds in (3.26). Since there is one single geodesic connecting $\bar{x}_0$ and y_0, this is a contradiction to (3.25).

From Varadhan's estimate [8]-[69]-[70], we know that

(3.27) $$p_s(\bar{x}_0, y_0) \geq e^{\frac{-d^2(\bar{x}_0, y_0) - \delta/2}{2s}}$$

From the estimate in [8]-VIII, Proposition 4.4, we know that

(3.28) $$p_s(x', y_0) \leq \frac{C}{s^N} e^{\frac{-d^2(x', y_0)}{2s}}$$

If $\bar{u}_0$ is such that $\pi \bar{u}_0 = \bar{x}_0$, if $\bar{u}^t_s = \psi^t_s(\sqrt{t}\, dw, \bar{u}_0)$, $\bar{x}^t_s = \pi \bar{u}^t_s$, we have

(3.29) $$\bar{Q}^t_{\bar{x}_0, y_0} (\sup_{0 \leq s \leq 1/2} |\sqrt{t}\, \beta_s - sn(\bar{x}_0)| \geq \frac{\varepsilon'}{2}) =$$

$$\frac{1}{p_t(\bar{x}_0, y_0)} E^P [1_{\sup_{0 \leq s \leq 1/2} |\sqrt{t}\, \beta_s - sn(\bar{x}_0)| \geq \frac{\varepsilon'}{2}}\ p_{t/2}(\bar{x}^t_{1/2}, y_0)]$$

We find that (3.29) is dominated by

$$
(3.30) \qquad \frac{Ce^{\frac{d^2(\bar{x}_0,y_0)+\delta/2}{2t}}}{t^N} E^P [\; 1 \sup_{0\le s\le 1/2} |\sqrt{t}\beta_s - sn(\bar{x}_0)| \ge \frac{\varepsilon'}{2} \; e^{\frac{-d^2(\bar{x}^t_{1/2},y_0)}{t}}]
$$

If D is the diameter of M, let h be a constant $\ge 2D + \delta$.

We now proceed as in Varadhan [69]-[70] and Azencott [6]. For $f \in L_2([0,\frac{1}{2}];R^n)$, set

$$
(3.31) \qquad I(f) = \int_0^{1/2} \frac{|f|^2}{2} \, ds
$$

Embed $L^2([0,1/2];R^n)$ in $\mathscr{C}^0([0,1/2];R^n)$ as after Definition 1.6, so that f will be identified with $\int_0^s f \, dv$. Similarly I is extended to $\mathscr{C}^0([0,1/2];R^n)$ by setting $I(e) = +\infty$ if $e \notin L^2([0,1/2];R^n)$. $\|\;\|$ is the norm in $\mathscr{C}^0([0,1/2];R^n)$.

From Theorem III.2.4 in Azencott [6], we know that since N is compact, for any $\rho > 0$, there is $\eta > 0$ such that for any $u'_0 \in N$, $f \in L_2([0,1/2];R^n)$ with $I(f) \le h^2$, if $u^t_s = \psi^t_s(\sqrt{t}\, dw, u'_0), \bar{x}^t_s = \pi u^t_s$, then

$$
(3.32) \qquad P\, [\| \sqrt{t}\, w_s - \int_0^s f \, dv \| \le \eta \; ; \; d(\bar{x}^f_{1/2}, \bar{x}^t_{1/2}) \ge \rho] \le e^{-\frac{h^2}{t}}
$$

Moreover $(I \le h^2)$ is $\mathscr{C}^0$-compact. Since I is l.s.c. on $\mathscr{C}^0([0,1/2];R^n)$, we may cover $(I \le h^2)$ by finitely many open balls $B(f_1,\eta'),\ldots B(f_k,\eta')$ such that $f_1,\ldots,f_k \in (I \le h^2)$, and moreover if $f \in B(f_i,\eta')$ then

(3.33) $I(f) \geq I(f_i) - \frac{\delta}{2}$

Of course we can suppose that $\eta' < \frac{\varepsilon'}{4} \wedge \eta$. Let $W^{\eta'}$ be the open set in $\mathscr{C}^0([0,1/2];R^n)$

(3.34) $W^{\eta'} = \bigcup_{1}^{k} B(f_i,\eta')$.

From [6], we know that

(3.35) $\overline{\lim}\ t\ \text{Log}\ P[\sqrt{t}\ w \in {}^c W^{\eta'}] \leq -h^2$

Moreover

(3.36) $E^P [1_{\sqrt{t}\ w \in W^{\eta'}}\ ;\ \|\sqrt{t}\ w_s - u_0'^{-1} sn(\bar{x}_0)\| \geq \frac{\varepsilon'}{2}\ e^{-\frac{d^2(\bar{x}_{1/2}^t, y_0)}{t}}]$

$$\leq \sum_{1}^{k} E^P [1_{\|\sqrt{t}\ w_s - u_0'^{-1} sn(\bar{x}_0)\| \geq \frac{\varepsilon'}{2}},\ \|\sqrt{t}\ w_s - \int_0^s f_i\ dv\| \leq \eta'\ e^{-\frac{d^2(\bar{x}_{1/2}^t, y_0)}{t}}]$$

Clearly each of the terms in the r.h.s. of (3.36) is $\neq 0$ only if :

(3.37) $\|\int_0^s f_i\ dv - u_0'^{-1}\ sn(\bar{x}_0)\| \geq \frac{\varepsilon'}{2} - \eta' \geq \frac{\varepsilon'}{4}$

Using (3.32) it is clear that if $\rho < h$

(3.38)
$$E^P [1_{\|\sqrt{t}\, w_s - u_0^{'-1} sn(\bar{x}_0)\| \geq \varepsilon'/2\ ;\ \|\sqrt{t}\, w_s - \int_0^s f_i\, dv\| \leq n'}$$

$$e^{-\frac{d^2(\bar{x}_{1/2}^t, y_0)}{t}}] \leq e^{\frac{-h^2}{t}} + e^{\frac{2h\rho}{t} - \frac{d^2(x_{1/2}^{f_i}, y_0)}{t}} P [\|\sqrt{t}\, w_s - \int_0^s f_i\, dv\| \leq n']$$

Using [6] again and (3.33), we know that

(3.39) $$\overline{\lim}\, t\, \text{Log}\, P\, [\|\sqrt{t}\, w - \int_0^1 f_i dv\| < n'] \leq -\inf_{f \in B(f_i, n)} I(f)$$

$$\leq - I(f_i) + \frac{\delta}{2}.$$

From (3.22), (3.23), (3.37), (3.39), we find that the non zero terms in (3.38) are dominated by

(3.40) $$e^{-\frac{h^2}{t}} + e^{\frac{3h\rho - \delta/2}{t} - \frac{d^2(\bar{x}_0, y_0)}{2t}}.$$

By choosing ρ small enough, we see easily from (3.30), (3.35), (3.40) that as $t \downarrow 0$, (3.30) is dominated by $e^{\frac{-\chi''}{t}}$, and so (3.19) is also dominated by $Ce^{\frac{-\chi''}{t}}$

Let $n'(\bar{x}_0)$ be the unique vector field in $T_{y_0} M$ such that $\exp_{y_0} sn'(\bar{x}_0)$ ($0 \leq s \leq 1$) is the unique minimizing geodesic connecting y_0 and $\bar{x}_0$ (which is still in $\tilde{U}^\varepsilon$). Using (2.42) and the boundedness of div b, we may show in the same way that

(3.41) $$\bar{Q}_{y_0, \bar{x}_0}^{'t} (\sup_{0 \leq s \leq 1/2} |\sqrt{t}\, \beta'_s + sn'(\bar{x}_0)| \geq \frac{\varepsilon'}{4}) \leq Ce^{\frac{-\chi''}{t}}$$

uniformly as $\bar{x}_0 \in \mathcal{V}^\varepsilon$. From Theorem 2.10, from the corollary of Theorem 2.11 and the fact that $-n'(\bar{x}_0)$ is the parallel translation of $n(\bar{x}_0)$ along the geodesic connecting $\bar{x}_0$ and y_0, we find that

(3.42) $\quad Q_{\bar{x}_0,y_0}^t \left(\sup_{1/2 \leq s \leq 1} |\sqrt{t}(\beta_1 - \beta_s) - (1-s)n(\bar{x}_0)| \geq \frac{\varepsilon'}{4} \right) \leq C e^{-\frac{\chi''}{t}}$

uniformly as $\bar{x}_0 \in \mathcal{V}^\varepsilon$. Now it is clear that :

(3.43) $\quad \left(\sup_{0 \leq s \leq 1} |\sqrt{t}\,\beta_s - sn(\bar{x}_0)| \geq \varepsilon' \right) \subset \left(\sup_{0 \leq s \leq 1/2} |\sqrt{t}\,\beta_s - sn(\bar{x}_0)| \geq \frac{\varepsilon'}{2} \right)$

$$\cup \left(\sup_{1/2 \leq s \leq 1} |\sqrt{t}(\beta_1 - \beta_s) - (1-s)n(\bar{x}_0)| \geq \frac{\varepsilon'}{4} \right)$$

(3.17) is then proved.

□

From Theorem 3.7, we now derive :

<u>Theorem 3.8</u> : As $t \downdownarrows 0$, $\dfrac{t \, \text{grad}_{\bar{x}_0} p_{t(1-s)}(\bar{x}_0, y_0)}{p_{t(1-s)}(\bar{x}_0, y_0)}$

converges to $\dfrac{n(\bar{x}_0)}{1-s}$ uniformly for $(s, \bar{x}_0) \in [0, 1/2] \times \mathcal{V}^\varepsilon$.

<u>Proof</u> : From Theorem 2.14 we have

(3.44) $\quad \dfrac{t \, \text{grad}_{\bar{x}_0} p_t(\bar{x}_0, y_0)}{p_t(\bar{x}_0, y_0)} = E^{Q_{\bar{x}_0,y_0}^t} \int_0^t \tilde{E}'_s \, \delta \beta_s$

$$= E^{Q_{\bar{x}_0,y_0}^t} [\tilde{E}'_t \beta_t - \int_0^t \frac{d\tilde{E}'_h}{dh} \beta_h \, dh]$$

Now E' can be calculated on $\bar{Q}^t_{x_0,y_0}$. Namely consider the differential equation :

(3.45) $\quad D\, E^t_s = t\, [-\frac{1}{2} S\, E^t_s + \nabla_{E^t_s} b]\, ds$

$E^t(0) = I$

where of course $D\, E^t_s$ is calculated on $\bar{x}^t_s$ ($0 \le s \le 1$).

If τ^0_s is the parallel translation operator along $\bar{x}^t_s$, set

(3.46) $\quad E'^t_s = [\tau^0_s]^{-1}\, E^t_s$

Since M is compact, S and $\nabla \cdot b$ are bounded. It is then clear that as $t \downarrow\downarrow 0$, $E'^t_s \to I$, $\dfrac{dE'^t_s}{ds} \to 0$ uniformly on $[0,1] \times \Omega$ and remain uniformly bounded. Moreover, (3.44) writes

(3.47) $\quad \dfrac{t\, \text{grad}_{\bar{x}_0}\, p_t(\bar{x}_0, y_0)}{p_t(\bar{x}_0, y_0)} = E^{\bar{Q}^t_{(\bar{x}_0,y_0)}}\, [\tilde{E}'^t_1 \sqrt{t}\beta_1 - \int_0^1 \dfrac{d\tilde{E}'^t_h}{dh} \sqrt{t}\beta_h\, dh]$.

From Theorem 3.7 and Proposition 2.12, we have that as $t \downarrow\downarrow 0$, the r.h.s. of (3.47) converges uniformly to $n(\bar{x}_0)$ for $\bar{x}_0 \in \mathcal{V}^\varepsilon$. Replacing t by $t(1-s)$, the Theorem follows. □

c) **Large deviations on flows.**

By using the results of b), we now establish the main technical result concerning large deviations on flows for the conditional measure $\bar{P}^t_{\bar{u}_0, y_0}$.

$h^1 \ldots h^n$ are n continuous functions defined on [0,1] with values in $T_{y_0} M$.

For $\bar{u}_0 \in N$, $q \in T^*_{y_0} M$, consider the stochastic differential equation on (Ω, P)

(3.48) $\quad du = tb'(u)ds + Y_i(u) [\sqrt{t}\, dw^i + <q,h^i> ds]$

$\quad\quad u(0) = \bar{u}_0$

Let $\psi^t(\sqrt{t}\, dw, q, .)$ be the flow of diffeomorphisms of N associated to (3.48). Of course, using the results of [10], we may and we will assume that $\psi^{\varepsilon^2}_s(\varepsilon dw, q, \bar{u}_0)$ is jointly continuous in $(\varepsilon, q, s, \bar{u}_0)$, C^∞ in $(\varepsilon, q, \bar{u}_0)$ with derivatives which are continuous in $(\varepsilon, q, s, \bar{u}_0)$.

Set

(3.49) $\quad h'^i_s = h^i_{1-s}$

We also consider the stochastic differential equation

$$du' = -tb(u')ds - Y_i(u')\,[\sqrt{t}\,dw'^i + <q,h'^i>\,ds]$$

$$u'(0) = \bar{u}'$$

and the flow $\psi'^t.(\sqrt{t}\,dw',q,.)$ which has the same properties as $\psi^t.(\sqrt{t}\,dw,q,.)$.

We first have a technical result.

Proposition 3.9 : For $\bar{u}_0 \in N$ consider the differential equation :

(3.50) $\quad dv^t = (\psi_s^t(\sqrt{t}\,dw,.)^{*-1} Y_i)(v^t) <q,h^i>\,ds$

$\quad\quad v^t(0) = \bar{u}_0$

Then P a.s, for any $(t,q,s,\bar{u}_0)$

(3.51) $\quad \psi_s^t(\sqrt{t}\,dw,q,\bar{u}_0) = \psi_s^t(\sqrt{t}\,dw,v_s^t)$

Proof : This is an obvious consequence of Theorem 4.1 in [11] and of the continuity in $(t,q,s,\bar{u}_0)$ of both sides of (3.51). □

Since under all the considered probability measures, w is a semi-martingale, all the previous results are still true for any of these probability measures, this by the results of Kunita [43].

$u_0 \in N$ such that $\pi u_0 = x_0$ is now fixed.

Definition 3.10 : λ is the vector of R^n

$$\lambda = u_0^{-1} \, n \, (x_0)$$

$\psi_s(\lambda,q,\bar{u}_0)$ is the solution of the differential equation

(3.52) $\quad du' = Y_i(u') \, [\lambda^i + <q,h'>] \, ds$

$\quad\quad\quad u'(0) = \bar{u}_0$

Definition 3.11 : For $m \in \mathbb{N}$, K^m is the set of functions defined on $\mathbb{N} \times \{q \in T^*_{y_0} M \, ; \, |q| \leq 1\}$ with values in $\mathbb{N}$ which have m continuous derivatives in (u,q).

When endowed with the topology of uniform convergence of the derivatives in (u,q) of order $\leq m$, K^m is a metrisable space. Let d^m be a distance on K^m.

In the sequel, $\psi_s^t(\sqrt{t} \, dw, q, \bar{u}_0)$, $\psi_s(\lambda, q, \bar{u}_0)$ will be considered as elements of K^m.

We now have the essential result :

Theorem 3.12 : For any $m \in \mathbb{N}$, $k \in \mathbb{N}$, $\eta > 0$, as $t \downarrow\downarrow 0$

(3.53) $\quad \bar{P}^t_{u_0, y_0} \, [\sup_{0 \leq s \leq 1} d^m(\psi_s^t(\sqrt{t} \, dw,.,.), \psi_s(\lambda,.,.)) \geq \eta] = o(t^k).$

Proof : We first prove that

(3.54) $\bar{P}^t_{u_0,y_0} [\sup_{0 \leq s \leq 1/2} d^m(\psi^t_s(\sqrt{t}\, dw, q, \bar{u}_0), \psi_s(\lambda, q, \bar{u}_0)) \geq \eta] = o(t^k)$

From Theorem 2.15, we know that under $\bar{P}^t_{u_0,y_0}$, a Brownian martingale $\bar{w}$ exists such that if $u^t_s = \psi^t_s(\sqrt{t}\, dw, u_0)$, $x^t_s = \pi u^t_s$, then

(3.55) $\sqrt{t}\, w_s = \sqrt{t}\, \bar{w}_s + t \int_0^s [u^t_h]^{-1} \dfrac{\text{grad}_x\, p_{t(1-h)}(x^t_h, y_0)}{p_{t(1-h)}(x^t_h, y_0)} dh$

So $\psi^t_s(\sqrt{t}\, dw,.,.)$ is associated to the stochastic differential equation

(3.56) $du' = tb'(u')ds + Y_i(u')(\sqrt{t}\, d\bar{w}^i +$

$+ t[[u^t_s]^{-1} \dfrac{\text{grad}_x\, p_{t(1-s)}(x^t_s, y_0)}{p_{t(1-s)}(x^t_s, y_0)}]^i\, ds + <q, h^i>\, ds)$

$u'(0) = \bar{u}_0$

We now use the transformation (3.50). Namely for $\bar{u}_0 \in N$, consider the differential equation

(3.57) $dv'^t = (\psi^{t*-1}_s(\sqrt{t}\, d\bar{w},..)Y_i)(v'^t)[t[[u^t_s]^{-1} \dfrac{\text{grad}_x\, p_{t(1-s)}(x^t_s, y_0)}{p_{t(1-s)}(x^t_s, y_0)}]^i$

$+ <q, h^i>\,]ds$

$v'^t(0) = \bar{u}_0$

Using again Theorem 4.1 in [11], we know that

$$(3.58) \quad \psi_s^t(\sqrt{t}\, dw, q, \bar{u}_0) = \psi_s^t(\sqrt{t}\, d\bar{w}, v_s'^t) \qquad 0 \leq s \leq 1/2$$

Of course, both sides are continuous in $(q, \bar{u}_0)$ so that except on a fixed P (or $\bar{P}_{u_0, y_0}^t$) negligible set in $F_{1/2}$, (3.58) holds for any s, q, $\bar{u}_0$.

Proposition 3.2 shows that for any $\varepsilon' > 0$, there is $\chi' > 0$ such that

$$(3.59) \quad \bar{P}_{u_0, y_0}^t [\sup_{0 \leq s \leq 1} d(x_s^t, \gamma_s) \geq \varepsilon')] \leq e^{\frac{-\chi'}{t}}$$

If $u_s = \psi_s(\lambda, 0, u_0)$, an obvious extension of the estimates (3.4)-(3.11) in the proof of Proposition 3.2 shows that for one $\chi' > 0$

$$(3.60) \quad \bar{P}_{u_0, y_0}^t [\sup_{0 \leq s \leq 1/2} d'(u_s^t, u_s) \geq \varepsilon'] \leq e^{\frac{-\chi'}{t}}$$

If $\bar{d}$ is any distance in TM, it is clear that for any $\delta > 0$, there is $\varepsilon' > 0$ such that $0 < \varepsilon' < \varepsilon$, and moreover if $d(x', \gamma_s) < \varepsilon'$ (where s is such that $0 \leq s \leq 1$), then $\bar{d}(n(x'), n(\gamma_s)) < \delta$.

By using (3.59)-(3.60) and Theorem 3.8, it is now clear that for any $\theta > 0$, there is $\chi'' > 0$ such that

$$(3.61) \quad \bar{P}_{u_0, y_0}^t [\sup_{0 \leq s \leq 1/2} |t[u_s^t]^{-1}\, \frac{\text{grad}_x\, p_{t(1-s)}(x_s^t, y_0)}{p_{t(1-s)}(x_s^t, y_0)}$$

$$- u_s^{-1}\, \frac{n(\gamma_s)}{1-s}| > \theta] \leq e^{\frac{-\chi''}{t}}$$

Now $\frac{n(\gamma_s)}{1-s} = \frac{d\gamma}{ds}$ is parallel along γ, so that

(3.62) $\quad \frac{u_s^{-1} n(\gamma_s)}{1-s} = \lambda$

Theorem I.2.1 in [10] shows that as $t \downarrow\downarrow 0$, if e is the identity mapping of N, and if d'^m is any distance in C^m (N,N), for any m, $k \in \mathbb{N}$, $\rho > 0$

(3.63) $\quad P[\sup_{0 \leq s \leq 1/2} d'^m(\psi_s^t(\sqrt{t}\, dw,.),e) \geq \rho] = o(t^k).$

From (3.57), (3.58), (3.61)-(3.63), (3.54) immediately follows.

To get (3.53), we must still be careful. Take $v_0 \in N$ such that $\pi v_0 = y_0$. Set

(3.64) $\quad y_s^t = x_{1-s}^t \qquad 0 \leq s \leq 1$

From Theorem 2.10, Theorem 2.15 and the Remark which follows, we know that under $\bar{P}_{u_0,y_0}^t$, y_s^t is a semi-martingale with respect to its own filtration. Let v_s^t be the parallel translation of v_0 along y_s^t. If $\sqrt{t}\, \beta'$ is the analogue of β' (constructed in Section 2) for y_s^t, if w' is defined by

(3.65) $\quad w_s' = v_0^{-1} \beta_s'$

v_s^t is the solution of :

(3.66) $\quad dv_s^t = -tb'(v_s^t)ds - \sqrt{t}\, Y_i(v_s^t).dw_s'^i$

$\qquad v_0^t = v_0.$

Let v_s be the parallel translation of v_0 along the geodesic γ_{1-s}. From (2.44) and (3.60), we find that

$$(3.67) \qquad \bar{P}^t_{u_0,y_0}[\sup_{0\leq s\leq 1/2} d'(v^t_s, v_s) \geq \varepsilon'] \leq e^{-\frac{\chi''}{t}}$$

From Theorem 2.11 and its Corollary, we find that $\bar{P}^t_{u_0,y_0}$ a.s. for $s \geq 1/2$

$$(3.68) \qquad u^t_s = v^t_{1-s} [v^t_{1/2}]^{-1} u^t_{1/2}$$

From (3.60), (3.67), (3.68) it is now obvious that for one $\chi'' > 0$

$$(3.69) \qquad \bar{P}^t_{u_0,y_0}[\sup_{0\leq s\leq 1} d'(u^t_s, u_s) \geq \varepsilon'] \leq e^{-\frac{\chi''}{t}}$$

Observe now that since $O(n)$ acts on R^n, if $a \in O(n)$, we may define the action of a on the vector of R^n ($<q,h'^1_s>,\ldots <q,h'^n_s>$) which we write $a <q,h'> .\psi'^t_{\cdot}(\sqrt{t}\, dw' + a <q,h'>\, ds,.)$ is now the flow associated to the stochastic differential equation :

$$(3.70) \qquad dv'' = -tb'(v'')ds - Y_i(v'')(\sqrt{t}\, dw'^i + (a <q,h'>)^i ds)$$

$$v''(0) = v''_0$$

By using the transformation (3.50) we may assume that $\psi'^t_s(\sqrt{t}\, dw' + a <q,h'>\, ds, v''_0)$ is jointly continuous in (t,q,a,s,v''_0).

We claim that $\bar{P}^t_{u_0,y_0}$ a.s. for any $s \geq 1/2$,

$\bar{u}_0 \in N, q \in T^*_{y_0} M,$

(3.71) $\quad \psi^t_s(\sqrt{t}\, dw, q, \bar{u}_0) = \{\psi'^t_{1-s}(\sqrt{t}\, dw' + v_0^{-1} u_1^t <q,h'>\, ds,.)$

$\circ\, [\psi'^t_{1/2}(\sqrt{t}\, dw' + v_0^{-1}u_1^t <q,h'>ds,.)]^{-1}\, (\psi^t_{1/2}(\sqrt{t}\, dw,q,\bar{u}_0)[u_1^t]^{-1} v_0)\} v_0^{-1} u_1^t$

Recall that $O(n)$ acts on N, so that say, the final product by $v_0^{-1} u_1^t$ makes sense.

By using (3.6^9), the analogue of (3.54) for $\psi'^t_s(\sqrt{t}\, dw' + a <q,h'>\, ds,.)$ and (3.71), (3.53) will then be proved.

We now establish (3.71). To simplify, we assume that $t=1$ and $b=0$, so that $p_s(x,y)$ is symmetric in (x,y) (the general case is left to the reader!). We use an idea very similar to what we did in the proof of Theorem 2.11.

Take ϵ, η with $0 < \epsilon < \eta < 1$. Since $\pi u_0 = x_0$, the law of $\phi_\epsilon(\omega, u_0)$ under P is given by

$$p_\epsilon(x_0, x)\, dx\, d\mu^x(u)$$

where $d\mu^x(u)$ is a probability measure in the fiber of N above x. Similarly the law of $\phi'_{1-\eta}(\omega, v_0)$ under P is given by:

(3.72) $\quad p_{1-\eta}(y_0, y)\, dy\, d\nu^y(v)$

On $N^3 \times \Omega$, we consider the probability measure

$$(3.73) \quad dS(\bar{u},u,v,\bar{\omega}) = \frac{p_\varepsilon(x_0,\pi\bar{u})p_{1-\eta}(\pi\phi_{\eta-\varepsilon}(\bar{\omega},\bar{u}),y_0)}{p_1(x_0,y_0)}$$

$$d\bar{u}\ d\mu^{\pi\bar{u}}(u)dv^{\pi\phi_{\eta-\varepsilon}(\bar{\omega},\bar{u})}(v)dP(\bar{\omega})$$

Let $\bar{w}_s$ be the trajectory of $\bar{\omega}$. Set

$$(3.74) \quad \bar{w}_s^{\varepsilon,\eta} = \bar{w}_{\eta-\varepsilon} - \bar{w}_{\eta-\varepsilon-s} \quad (0 \le s \le \eta - \varepsilon)$$

Let $\{\mathcal{G}_s\}_{s\ge 0}$ be the filtration associated to

$$(3.75) \quad \mathcal{G}_s = \mathcal{B}(\bar{u},u) \vee \mathcal{B}(\bar{w}_h \mid h \le s)$$

dS being absolutely continuous with respect to $d\bar{u}\ d\mu^{\pi\bar{u}}(u)dP(\bar{\omega})$, $\bar{w}_s$ is a $\{\mathcal{G}_s\}_{s\ge 0}$-semi-martingale.

Set

$$(3.76) \quad w_s = u^{-1}\ \bar{u}\ \bar{w}_s$$

Of course $u^{-1}\bar{u}$ makes sense since $\bar{u}$ and u are in the same fiber. Moreover w_s is still a $\{\mathcal{G}_s\}_{s\ge 0}$-semi-martingale under S and its law is equivalent to the Brownian measure. Now we claim that S a.s., for any q

$$(3.77) \quad \psi^1_\cdot(dw,q,.) = \psi^1_\cdot(d\bar{w} + \bar{u}^{-1}u <q,h> ds,.u^{-1}\bar{u})\bar{u}^{-1}u$$

Indeed the action of both sides of (3.77) on $u' \in N$ are a.s. identified (this follows from the equation of the connection (2.2)) and so (3.77) follows from the continuity of both sides of (3.77) in all the considered variables.

By (2.57) we know that the Jacobian of the mapping $\bar{u} \to \phi_{\eta-\varepsilon}(\bar{\omega},\bar{u}) = \bar{v}$ is 1, so that from (2.51), in the new variables $(\bar{v},u,v,\bar{\omega}^{\varepsilon,\eta})$, S writes:

$$(3.78) \quad dS'(\bar{v},u,v,\bar{\omega}^{\varepsilon,\eta}) = \frac{p_\varepsilon(x_0,\pi\phi'_{\eta-\varepsilon}(\bar{\omega}^{\varepsilon,\eta},\bar{v}))p_{1-\eta}(\pi\bar{v},y_0)}{p_1(x_0,y_0)}$$

$$d\bar{v} \, d\mu^{\pi\phi'_{\eta-\varepsilon}(\bar{\omega}^{\varepsilon,\eta},\bar{v})}(u) \, dv \, ^{\pi\bar{v}}(v) \, dP(\bar{\omega}^{\varepsilon,\eta}).$$

If $\{\bar{\mathscr{G}}'_s\}_{s\geq 0}$ is the filtration associated to

$$(3.79) \quad \bar{\mathscr{G}}'_s = \mathscr{B}(\bar{v},v) \vee \mathscr{B}(\bar{w}_h^{\varepsilon,\eta}|h\leq s)$$

and if

$$(3.80) \quad w'_s = \bar{v}^{-1} \bar{w}^{\varepsilon,\eta}_s$$

we still find that $\bar{w}^{\varepsilon,\eta}_s$, w'_s are $\{\bar{\mathscr{G}}'_s\}_{s\geq 0}$ semi-martingales and moreover S a.s., for any $q \in T^*_{y_0} M$, $a \in O(n)$

$$(3.81) \quad \psi'^{1}(dw' + a <q,h'>,.) = \psi'^{1}(d\bar{w}^{\varepsilon,\eta} + \bar{v}^{-1} \, va <q',h>,..v^{-1}\bar{v})\bar{v}^{-1}v$$

Now under S, the law of $\bar{w}_s$ ($0\leq s\leq \eta-\varepsilon$) is equivalent to the Brownian measure P.

By [10] Theorem I.3.1, we find that S a.s, for any $s(0\leq s\leq \eta-\varepsilon), q \in T^*_{y_0}M, a \in O(n)$

(3.82) $\quad \psi^1_s(d\bar{w} + a<q,h>,.) = \psi'^1_{\eta-\varepsilon-s}(d\bar{w}^{\varepsilon,\eta} + a<q,h'>,.) \circ \psi^1_{\eta-\varepsilon}(d\bar{w} + a<q,h>,.)$

From (3.77), (3.81), (3.82) we see that S a.s, for any $s(0\leq s\leq \eta-\varepsilon)$, $q \in T^*_{y_0}M$, $u' \in N$

(3.83) $\quad \psi^1_s(dw,q,u') = \psi'^1_{\eta-\varepsilon-s}(dw' + v^{-1}\bar{v}\ \bar{u}^{-1}u <q,h'>,$

$$\psi^1_{\eta-\varepsilon}(dw,q,u')u^{-1}\ \bar{u}\ \bar{v}^{-1}v)v^{-1}\ \bar{v}\ \bar{u}^{-1}u$$

Now $\bar{v}\ \bar{u}^{-1}u$ is exactly the parallel translation $u_{\eta-\varepsilon}$ of u along the $\{\mathcal{G}_s\}_{s\geq 0}$ semi-martingale $x_s = \pi\varphi_s(\bar{\omega},\bar{u})$, so that (3.83) reads

(3.84) $\quad \psi^1_s(dw,q,u') = \psi'^1_{\eta-\varepsilon-s}(dw' + v^{-1}u_{\eta-\varepsilon}<q,h'>,\ \psi^1_{\eta-\varepsilon}(dw,q,u')u_{\eta-\varepsilon}^{-1}v)v^{-1}u_{\eta-\varepsilon}$

From (3.84) and some obvious identifications of probability laws which are left to the reader, it is not difficult to deduce (3.71).

When there is a drift b, formula (3.73) has to be slightly modified in an obvious way.

□

<u>Remark 1</u> : The difficulty in the proof of Theorem 3.12 is that under $P^1_{u_0,y_0}$, when time reversing the process $\psi^1_s(dw,u_0)$, we do not obtain the process $\psi'^1_s(dw',v_0)$.

IV. TAYLOR EXPANSION OF THE CONDITIONAL PROBABILITY : THE ELLIPTIC CASE

In this section, we will use the results of Section 3 to expand the conditional probability $\bar{P}^t_{u_0,y_0}$ as a Taylor series, and so find a Taylor expansion of $p_t(x_0,y_0)$ in terms of stochastic integrals with respect to a Brownian bridge.

As we have seen in the Introduction, we will parametrize the Hilbert submanifold $K'^{u_0}_{y_0}$ of $H = L_2([0,1]; R^n)$ given by :

$$K'^{u_0}_{y_0} = \{h \in H \; ; \; \pi \varphi_1(hds, u_0) = y_0\}$$

using the split $H = H_1 \oplus H_2$, where H_1 is equal to $T_\lambda K'^{u_0}_{y_0}$ and H_2 is the finite dimensional $[T_\lambda K'^{u_0}_{y_0}]^\perp$. Of course such a parametrization is only valid locally, in the same way as in Theorem 1.20.

This permits us to describe the conditional law $\bar{P}^t_{u_0,y_0}$ on a neighborhood of λ in terms of the Gaussian measure P_1 on H_1, since on this neighborhood the component v_2 is a function of v_1. Of course, this is considerably complicated by the fact that H_1 has 0 measure for P_1, $K'^{u_0}_{y_0}$ 0 measure for $\bar{P}^t_{u_0,y_0}$, so that $\bar{P}^t_{u_0,y_0}$ is in fact a cylindrical measure on the Hilbert manifold $K'^{u_0}_{y_0}$. Although this concept can appear of formidable complexity, the fact that $K'^{u_0}_{y_0}$ has finite codimension and that we use stochastic flows makes the difficulty disappear.

Of course such a parametrization is only local while we need a global control of $\overline{P}^t_{u_0,y_0}$ to be sure that indeed everything really concentrates around λ.

In a), we describe a first split of H, and in b) the corresponding split of the Brownian measure P as a product $P_1 \otimes P_2$. In c), we express P_1 in terms of a standard Brownian bridge. In d), we use the results of Section 3 to establish large deviation results for flows contructed for the measure P_1. In e), we describe a local change of variables so that instead of $(w^1, w^2) = w$, we now use $(w^1, \pi \psi^t_1 (\sqrt{t}dw, u_0))$. In f), a global asymptotic expression of $p_t(x_0,y_0)$ is given as a Laplace integral on (Ω, P_1), which makes appear the energy functional, and the Malliavin covariance matrix of the problem.

In g), extending a result of Elworthy and Truman [29], we express the determinant of the Jacobian of the exponential mapping on a Riemannian manifold as a path integral over a Brownian bridge. In h) the result of Molchanov [54] which gives an equivalent for $p_t(x_0,y_0)$ as $t \downarrow 0$, is obtained. In the course of the proof of this result, some key estimates are established, which will be essent- in obtaining the whole Taylor expansion of $p_t(x_0,y_0)$. As pointed out in the Introduction, we cannot use the results of Schilder [57], and this is the basic reason why we do need such estimates. In i), the Taylor expansion of $p_t(x_0,y_0)$ in terms of expectations of stochastic integrals is obtained. In j) another split of H is briefly discribed, which simplifies some computations. In k), a method is described to obtain explicitly the Minakshishundaram-Pleijel expansion of $p_t(x_0,x_0)$.

Of course all this section is much connected in spirit with the mathematical physics litterature on path integrals. We have already mentioned Schilder [57]. The pioneering work of De Witt Morette [19], [20], De Witt Morette, Maheswari, Nelson [21] has also been very inspiring, as well as Langouche and al [81].

The articles by Truman [67], [68], Elworthy-Truman [29], Davies-Truman [17] are also very relevant. In [17], [67], [68], the authors consider the case of a quasiclassical expansion for the Euclidean Laplacian plus a potential, so as to obtain the asymptotic expansion as $t \downarrow\downarrow 0$ of $\int \exp - \frac{S_0(y)}{t} T_0(y) \, p_t(x_0,y) \, dy$ and of $\int p_t(x,x) \, dx$. In [29], Elworthy and Truman have considered the problem of expanding $\int e \frac{-S_0(y)}{t} T_0(y) \, p_t(x_0,y) \, dy$ on a Riemannian manifold. Although the computations in [29] are closely related to what is done here, the main difficulty in our paper is that we are in fact working on a conditional probability law, i.e. on a submanifold of the path space. This creates several technical difficulties.

Of course Molchanov [54] has indicated the theoretical possibility of obtaining an expansion of $p_t(x_0,y_0)$ using probabilistic methods.

Let us again point out that Azencott [7] had indicated the possibility of expanding $p_t(x_0,y_0)$ in terms of stochastic integrals. The main difference is that our computations are entirely intrinsic and exhibit the key importance of the objects introduced by Malliavin [46],[47].

a) **A first split of the Hilbert space H**

As in Section 1, H is the Hilbert space $L_2([0,1]) ; R^n)$.

$u_0 \in N$ such that $\pi u_0 = x_0$ is fixed as after Definition 3.10, in which λ has been defined.

$f_s(.)$ denotes the flow of diffeomorphisms of N which is associated to the differential equation

(4.1) $\qquad dv = Y_i(v)\lambda^i d \; ; v(0) = v_0$

u_s is defined to be equal to $f_s(u_0)$. u_s is the parallel translation of u_0 along γ.

We first define the functions $h^1 \ldots h^n$.

Definition 4.1. : For $1 \le i \le n$, h_s^i is the continuous function defined on $[0,1]$ with values in $T_{y_0} M$

$$h_s^i = \pi^* f_1^* (f_s^{*-1} Y_i)(u_0)$$

In the sequel, we will use the results of Section 3 with precisely this choice of $h^1 \ldots h^n$.

We now modify (for obvious reasons) the definition of H_1 and H_2 given in Definition 1.13.

Definition 4.2. : H_1 is the set of $v \in H$ such that

(4.2) $\quad \pi^* f_1^* \int_0^1 (f_s^{*-1} Y_i)(u_0) v^i \, ds = 0$

H_2 is the finite dimensional subspace of H which is the image of $T_{y_0}^* M$ by the mapping ρ

(4.3) $\quad q \in T_{y_0}^* M \to \rho(q) = (<q, h_s^1>, \ldots <q, h_s^n>).$

We finally define the associated Malliavin covariance matrix.

Definition 4.3 : C is the linear mapping from $T_{y_0}^* M$ into $T_{y_0} M$ given by :

(4.4) $\quad q \in T_{y_0}^* M \to \pi^* f_1^* \int_0^1 (f_s^{*-1} Y_i)(u_0) <q, \pi^* f_1^* (f_s^{*-1} Y_i)(u_0)> \, ds.$

We then have the obvious analogue of Theorem 1.19.

Theorem 4.4. : The matrix C is invertible. H_1 and H_2 are orthogonal subspaces of H and moreover

(4.5) $\quad H = H_1 \oplus H_2$

If $v \in H$, the orthogonal projection $\bar{P}_2 v$ of v on H_2 is given by :

(4.6) $\quad (\bar{P}_2 v)_t = \rho(q(v))$

where

(4.7) $\quad q(v) = C^{-1} \pi^* f_1^* \int_0^1 (f_s^{*-1} Y_i)(u_0) v^i \, ds$

- 122 -

Proof : The invertibility of C is obvious from the fact that for any $u \in N$, $\pi^* Y_1(u), \ldots \pi^* Y_n(u)$ span $T_{\pi u} M$. The remaining part of the Theorem is trivial. □

Remark 1 : The projection operator $\overline{P}_2$ is not especially pleasant. This is the main reason why in j), we show how to split H in another way so that when $T^*_{y_0} M$ and $T_{y_0} M$ are identified, the corresponding matrix C' is the identity.

We now give a covariant characterization of H_1 and H_2.

Definition 4.5 : For $v \in H$, consider the differential system calculated along the geodesic γ

(4.8) $\quad \dfrac{DJ}{Ds} = u_s v + J' \quad ; \quad J(0) = 0$

$\dfrac{DJ'}{Ds} = R(\dfrac{d\gamma}{ds}, J) \dfrac{d\gamma}{ds} \quad ; \quad J'(0) = 0$

k is the linear mapping

(4.9) $v \in H \to k v = J_1 \in T_{y_0} M$.

For $X \in T_{y_0} M$, let $\overline{J}$ be the Jacobi field

(4.10) $\quad \dfrac{D^2 \overline{J}}{Ds^2} + R(\overline{J}, \dfrac{d\gamma}{ds}) \dfrac{d\gamma}{ds} = 0$

$\overline{J}(1) = 0, \quad \dfrac{D\overline{J}_1}{Ds} = X$

k^* is the linear mapping

(4.11) $\quad X \in T_{y_0} M \to k^* v = u_s^{-1} \dfrac{DJ_s}{Ds} \in H$

As the notation suggests, k and k^* are adjoint to each other.

<u>Theorem 4.6</u> : H_1 is the set of $v \in H$ such that

(4.12) $\quad k(v) = 0$

H_2 is the set of $v \in H$ such that $X \in T_{y_0} M$ exists for which

(4.13) $\quad v = k^*(X)$

The operator $k\, k^*$ which sends $T_{y_0} M$ into itself is invertible and coincides with C (when $T_{y_0}^* M$ and $T_{y_0} M$ are identified). Morever

(4.14) $\quad \overline{P}_2 = k^*(k\, k^*)^{-1} k$

<u>Proof</u> : Using equation (2.2), it is easy to see that H_1 consists of the $v \in H$ such that if θ, ω is the solution of

(4.15) $\quad \dfrac{d\theta}{ds} = v + \omega \lambda \quad ; \theta(0) = 0$

$ \dfrac{d\omega}{ds} = \Omega(\lambda, \theta) \quad ; \omega(0) = 0$

then $\theta(1) = 0$. If $J_s = u_s\, \theta_s$, $J'_s = u_s\, \omega_s \lambda$, (J, J') is the solution of (4.8), and so (4.12) is proved. The proof of the theorem is now easy. □

Remark 2 : A key observation is that $\lambda \in H_2$. The unique $q \in T^*_{y_0} M$ such that $\lambda = \rho(q)$ is given by

(4.16) $q = \frac{d\gamma}{ds} 1$

Similarly the unique $X \in T_{y_0} M$ such that $\lambda = k^*(X)$ is also

(4.17) $X = \frac{d\gamma}{ds} 1$

Because of (4.16), and using the definition of h given in (4.1), we will write $\psi_s(\frac{d\gamma}{ds}1 + q, u_0)$ instead of $\psi_s(\lambda, q, u_0)$ (which was defined in Definition 3.10).

b) <u>The split of the Brownian measure</u>

With the orthogonal split $H = H_1 \oplus H_2$, we now associate the corresponding split of the Wiener measure P, considered as a cylindrical measure on H.

<u>Definition 4.7</u> : P_1 is the canonical Gaussian cylindrical measure on the Hilbert space H_1.

P_2 is the standard Gaussian measure on the finite dimensional Hilbert space H_2.

Recall that the space $\mathscr{C}^0([0,1] ; R^n)$ has been defined in Definition 1.6. The following result shows that in the same way as P, P_1 defines a probability measure on $\mathscr{C}^0([0,1] ; R^n)$.

Theorem 4.8: For $w \in \mathcal{C}^0([0,1]; R^n)$, consider the system

(4.18) $J_t = u_t w_t + \int_0^t J'_s \, ds$

$J'_t = \int_0^t R(\frac{dy}{ds}, J) \frac{dy}{ds} \, ds.$

Set

(4.19) $\dot{w}^2 = k^*(k\,k^*)^{-1} J_1$

$w_t^1 = w_t - \int_0^t \dot{w}_s^2 \, ds$

Under P, w^1 and $\dot{w}^2$ are independent. Moreover the law of dw^1 is P_1, and the law of $\dot{w}^2$ is P_2.

Proof: We give a rough proof, which can of course be made correct. $\dot{w}^2$ is the projection of dw on H_2, and so dw^1 is the projection of dw on H_1. The result follows. □

Note that if $a_t = u_t^{-1} J_t$, (4.18) writes in the following equivalent form

(4.20) $a_t = w_t + \int_0^t \alpha_\lambda \, ds$

$\alpha_t = \int_0^t \Omega(\lambda, a) \, ds$

Corollary: In Theorem 4.8, we have

(4.21) $J_1(dw) = \pi^* f_1^* \int_0^1 (f_s^{*-1} Y_i)(u_0) \cdot dw^i$

$(\dot{w}_s^2)^i = \langle C^{-1} J_1(dw), h_s^i \rangle$

Proof : This is trivial from the results of a), and in particular from the proof of Theorem 4.6. □

Remark 3 : (4.18) - (4.20) makes sense for any $w \in \mathcal{C}^0([0,1] ; \mathbb{R}^n)$ so that there is no probability involved in solving (4.18). Similarly, by integrating by parts, we have :

(4.22) $\quad J_1(dw) = \pi_1^* f_1^* \left[(f_1^{*-1} Y_i)(u_0) w_1^i - \int_0^1 f_s^{*-1} [Y_j, Y_i] \lambda^j w_s^i \, ds \right]$

so that $J_1(dw)$ is as well defined in (4.21) for any $w \in \mathcal{C}^0([0,1] ; \mathbb{R}^n)$. This fact is <u>essential</u> in the sequel.

Remark 4 : As we will see in Section 4 c), under P_1, w^1 is a semi-martingale on [0,1], so that it is possible to integrate **predictable** processes with respect to w^1.

Moreover one can show that for $s < 1$, P_1 and P are equivalent on F_s.

Remark 5 : Contrary to dw (and dw^1), w^2 is indeed an element of H, which has even C^∞ trajectories.

c) **Expression of P_1 in terms of a standard Brownian bridge**

Later on, we will use another useful description of P_1.

Definition 4.9: Q denotes the probability measure on Ω of the R^n-valued Brownian bridge $a_t (0 \le t \le 1)$, with $a_0 = a_1 = 0$.

Under P, the law of $w_t - t w_1$ is exactly Q.

We now claim.

Theorem 4.10: On (Ω, Q), consider the differential system

$$(4.23) \qquad \alpha_t = \int_0^t \Omega_{u_s}(\lambda, a_s)\, ds,$$

$$w_t^1 = a_t - \int_0^t \alpha_s \lambda\, ds.$$

If K' is given by

$$K' = E^Q \exp\{ \int_0^1 <\Omega(a_s, \lambda)\lambda, a_s> ds - \frac{1}{2} \int_0^1 |\alpha_s \lambda|^2 ds \}$$

then K' is $< +\infty$. If Q' is the probability measure

$$(4.24) \qquad dQ'(a) = \frac{\exp\{ \int_0^1 <\Omega(a_s, \lambda)\lambda, a_s> ds - \frac{1}{2}\int_0^1 |\alpha_s \lambda|^2 ds\}\, dQ(a)}{K'}$$

under Q', the law of w^1 is P_1.

Proof : We give a short proof. Assume first that the law of a is P. Then if M_t is given by :

$$M_t = \exp\{\int_0^t <\alpha\lambda, \delta a> - \frac{1}{2} |\alpha\lambda|^2 \, ds\}$$

M_t is a local martingale. By stopping M adequately and using the properties of the Girsanov transformation, it is fairly easy to check that M_t is a martingale. Obviously

$$M_1 = \exp\{<\alpha\lambda, a_1> + \int_0^1 <\Omega(a,\lambda)\lambda, a> \, ds - \frac{1}{2}\int_0^1 |\alpha\lambda|^2 \, ds\}$$

If $Q^{a'}$ is the law of the Brownian bridge with $a(0) = 0$, $a(1) = a'$, $Q^{a'}$ depends continuously on a' so that if g(a') is given by

$$g(a') = E^{Q^{a'}} M_1$$

g is l.s.c. If $dP' = M_1 \, dP$, under the probability measure P', a_1 has a non degenerate gaussian law r(x) dx, where $r \in C_b^\infty(R^n)$. Obviously

(4.25) $\quad r(x) = \dfrac{g(x)}{(\sqrt{2\pi})^n} e^{-\frac{|x|^2}{2}}$ a.e

We can then find $x_k \to 0$ such that equality in (4.25) holds at x_k. Since g is l.s.c., $g(0) < +\infty$.

The Theorem is now obvious using (4.20) and Theorem 4.8. □

d) Large deviations for w^1

As indicated in the Introduction of the paper, we now must prove that localization is indeed possible in the evaluation of $p_t(x_0, y_0)$.

Note that by using Theorem 4.8, we know that w^1 is well defined on (Ω, P). Moreover since

$$(\dot{w}^2)^i = < C^{-1} J_1(dw), h^i >$$

it is feasible to define $\psi_s^t(\sqrt{t}dw^1, q, .)$ by the formula

(4.26) $\quad \psi_s^t(\sqrt{t} dw^1, q, .) = \psi_s^t(\sqrt{t}dw, q - \sqrt{t}C^{-1} J_1(dw), .)$

Of course (4.26) is not purely formal. Namely by using the results of Jeulin [38], it can be easily proved that w_s ($0 \le s < 1$) is still a semi-martingale with respect to the enlarged filtration associated to the σ-fields

$$B(w_h | h \le s) \vee \mathcal{B}(J_1(dw))$$

w_s^1 ($0 \le s < 1$) is then a semi-martingale with respect to this enlarged filtration, so that (4.26) is an identity for $s < 1$. As pointed out in Remark 4, the semi-martingale property of w, w^1 is in fact true on $[0,1]$ so that, under P, (4.26) holds on $[0,1]$.

Since $\bar{P}^t_{u_0, y_0}$ is equivalent to P on each F_s ($0 \le s < 1$), (4.26) makes sense under $\bar{P}^t_{u_0, y_0}$ for $s < 1$. From Theorem 2.15, w is a semi-martingale on $[0,1]$ under $\bar{P}^t_{u_0, y_0}$, and so the r.h.s. of (4.26) makes sense on $[0,1]$. The l.h.s. of (4.26) clearly extends by continuity on $[0,1]$.

We now have the key result.

Theorem 4.11 : For any $k \in N$, $\delta > 0$, as $t \downarrow\downarrow 0$

(4.27) $\bar{P}^t_{u_0, y_0} [(\sup_{0 \leq s \leq 1} |\sqrt{t}\, w^1_s| \geq \delta) \cup (\sup_{0 \leq s \leq 1} |\sqrt{t}\, \dot{w}^2_s - \lambda| \geq \delta)] = o(t^k)$.

For any $\delta > 0$, $k \in N$, $m \in N$

(4.28) $\bar{P}^t_{u_0, y_0} [\sup_{\substack{0 \leq s \leq 1 \\ t' \leq t}} d^m(\psi^{t'}_s(\sqrt{t'}\, dw^1, \frac{d\gamma}{ds}1 + \cdot, \cdot), \psi_s(\frac{d\gamma}{ds}1 + \cdot, \cdot)) \geq \delta] = o(t^k)$

Proof : From (4.16), (4.22), Theorem 3.7, and (4.21), it is clear that as $t \downarrow\downarrow 0$

(4.29) $\bar{P}^t_{u_0, y_0} [\sup_{0 \leq s \leq 1} |\sqrt{t}\, \dot{w}^2_s - \lambda| \geq \delta] \leq e - \frac{\chi''}{t}$

Since

$$w_t = w_t^1 + \int_0^t \dot{w}_s^2 \, ds$$

by using Theorem 3.7 again, (4.27) follows easily.

We now prove (4.28). We use the same notations as in the proof of Theorem 3.12. From (4.26), we know that for $\bar{u}_0 \in N$

(4.30) $\quad \psi_s^{t'}(\sqrt{t}\, dw^1, \frac{d\gamma}{ds}1 + q, \bar{u}_0) = \psi_s^{t'}(\sqrt{t'}\, dw, \frac{d\gamma}{ds}1 - C^{-1} J_1(\sqrt{t'}\, dw)$

$\quad\quad\quad + q, \bar{u}_0)$

Of course (4.30) is verified $\bar{P}_{u_0, y_0}^t$ a.s. for any $(s, t', q, \bar{u}_0)$ with $s < 1$, since both sides of (4.30) are jointly continuous in all the considered variables. Using (3.55), (4.30) is also equal to:

(4.31) $\quad \psi_s^{t'}(\sqrt{t'}\, d\bar{w} + \left[\frac{t}{t}\right]^{1/2} \left[u_s^t\right]^{-1} \frac{\text{tgrad}_x \, P_{t(1-s)}(x_t^s, y_0)}{P_{t(1-s)}(x_s^t, y_0)} \, ds,$

$\frac{d\gamma}{ds}1 - C^{-1} J_1(\sqrt{t'}\, dw) + q, \bar{u}_0)$

Instead of (3.57), we consider the differential equation

(4.32) $\quad dv_s^{t,t'} = (\psi_s^{t'*-1}(\sqrt{t'}\, d\bar{w}, .) Y_i)(v_s^{t,t'})\{\left[\frac{t}{t}\right]^{1/2}\left[\left[u_s^t\right]^{-1}\right.$

$\left.\frac{\text{tgrad}_x \, P_{t(1-s)}(x_t^s, y_0)}{P_{t(1-s)}(x_s^t, y_0)}\right]^i + <\frac{d\gamma}{ds}1 - C^{-1} J_1(\sqrt{t'}\, dw) + q, h_s^i> \} \, ds.$

$v^{t,t'}(0) = \bar{u}_0$

We know that as in (3.58), (4.31) is equal to

(4.33) $\psi_s^{t'}(\sqrt{t'}\, d\bar{w}, v_s^{t,t'})$

Of course, since $v_s^{t,t'}$ depends continuously on $(t',s,q,\bar{u}_0)$, the equality between (4.31) and (4.33) holds $\bar{P}_{u_0,y_0}$ a.s. for any $(t',s, q, \bar{u}_0)$ with $s < 1$.

From Theorem I.2.1. in [10], we know that for any $\rho > 0$, $k \in \mathbb{N}$

(4.34) $P\,[\sup_{\substack{0 \le s \le 1/2 \\ 0 \le t' \le t}} d'^m\,[\psi_s^{t'}(\sqrt{t'}\,dw,.), e] \ge \rho] = o\,(t^k)$

(this is a stronger statement than (3.63)). By using Remark 2, we know that if

(4.35) $J_1(\lambda) = \pi^* \, f_1^* \int_0^1 (f_s^{*-1}\, Y_i)\,(u_0)\, \lambda^i\, ds$

then

(4.36) $J_1(\lambda) = \dfrac{d\gamma_1}{ds}$

$C^{-1}\, J_1(\lambda) = \dfrac{d\gamma_1}{ds}$

From Remark 2, we know that

(4.37) $<\dfrac{d\gamma}{ds}1, h_s^i> = \lambda^i = \left[u_s^{-1}\dfrac{d\gamma}{ds}\right]^i$

(4.22) and (4.27) show that

(4.38) $\bar{P}^t_{u_0,y_0} [|\sqrt{t} \, J_1(dw) - \frac{d\gamma}{ds}1| \geq \delta] = o(t^k)$

Finally if $t' \leq t$

(4.39) $\left[\frac{t'}{t}\right]^{1/2} | [u^t_s]^{-1} \frac{\text{tgrad}_x \, P_{t(1-s)}(x^s_t, y_0)}{P_{t(1-s)}(x^s_t, y_0)} - \lambda | \leq$

$| [u^t_s]^{-1} \frac{\text{tgrad}_x \, P_{t(1-s)}(x^s_t, y_0)}{P_{t(1-s)}(x^s_t, y_0)} - \lambda |$

By using (4.32), (4.34), (4.38), (4.39), it is now easy to proceed as in the proof of Theorem 3.12, so that

(4.40) $\bar{P}^t_{u_0,y_0} [\sup_{\substack{0 \leq s \leq 1/2 \\ 0 \leq t' \leq t}} d^m(\psi^{t'}_s(\sqrt{t'} \, dw^1, \frac{d\gamma}{ds}1 + .,.), \psi_s(\frac{d\gamma}{ds}1+.,.)) \geq \delta] = o(t^k)$.

We still use the notations of the proof of Theorem 3.12. We here take $v_0 = f_1(u_0)$. w' is defined as in (3.65), so that

(4.41) $w'_s = v_0^{-1} u^t_1 (w_1 - w_{1-s})$.

Using Theorem 2.11, its Corollary and the proof of Theorem 3.12, it is crucial to note that under $\bar{P}^t_{u_0,y_0}$ w'_s plays the same role for the process

$y_s^t = \pi u_{1-s}^t$ as w_s for the process $x_s^t = \pi u_s^t$. In particular under $\overline{P}_{u_0,y_0}^t$, w' is a semi-martingale.

For $t' \leq t$, consider the stochastic differential equation

(4.42) $\quad d\,\overline{u} = -\,t'b'(\overline{u})ds - Y_i(\overline{u})(\sqrt{t'}\,dw'^i + (v_0^{-1}\,u_1^t$

$$< \frac{d\gamma}{ds}1 + q - C^{-1}\,J_1(\sqrt{t'}\,dw), h' >)^j\,ds)$$

$\overline{u}(0) = \overline{u}_0$.

Of course the transformation (3.50) is still used on (4.42) so that although the anticipating u_1^t appears, the solution of (4.42) is still well defined. Let $\overline{\psi}_s^{t,t'}(\sqrt{t'}\,dw', \frac{d\gamma}{ds}1 + q - C^{-1}\,J_1(\sqrt{t'}dw),\overline{u}_0)$ be the unique solution of (4.42). Of course $\overline{\psi}_\cdot^{t,t'}$ is jointly continuous in $(s,t',q,\overline{u}_0)$. If $b = 0$, under $\overline{P}_{u_0,y_0}^t$, we may write a decomposition of $\sqrt{t}\,w'$ very similar to (3.55). Using (3.65), and the fact that $u_1 = v_0$, it is then easy to proceed as before and conclude that

(4.43) $\quad \overline{P}_{u_0,y_0}^t \left[\sup_{\substack{0 \leq s \leq 1/2 \\ 0 \leq t' \leq t}} d^m [\overline{\psi}_s^{t,t'}(\sqrt{t'}\,dw', \frac{d\gamma}{ds}1 - C^{-1}\,J_1(\sqrt{t'}\,dw)$

$$+.,.), \psi_{1-s} \circ \psi_1^{-1} (\frac{d\gamma}{ds}1+.,.)] \geq \delta \right] = o(t^k).$$

If $b \neq 0$, we may use (2.44) and still obtain (4.43). Now by proceeding as for the proof of (3.71), it is not difficult to see that if $s \geq 1/2$

$$(4.44) \quad \psi_s^{t'}(\sqrt{t'}\, dw, \tfrac{d\gamma}{ds}1 - C^{-1} J_1(\sqrt{t'}\, dw) + q, \bar{u}_0) =$$

$$\left\{ \left(\bar{\psi}_{1-s}^{t,t'} \circ \left[\bar{\psi}_{1/2}^{t,t'} \right]^{-1} \right) \!\left(\sqrt{t'}\, dw', \tfrac{d\gamma}{ds}1 - C^{-1} J_1(\sqrt{t'}\, dw) + q, . \right) \right.$$

$$\left. \left(\psi_{1/2}^{t'} \!\left(\sqrt{t'}\, dw, \tfrac{d\gamma}{ds}1 - C^{-1} J_1(\sqrt{t'}\, dw) + q, \bar{u}_0 \right)\!(u_1^t)^{-1} v_o \right) \right\} v_o^{-1} u_1^t \, .$$

From (3.69), (4.40), (4.43), (4.44), it is now easy to obtain (4.28). The Theorem is proved. □

e) A local change of variables

The key property of the split $H = H_1 \oplus H_2$ is the trivial relation

$$(4.45) \quad \pi^* \frac{\partial \psi_1}{\partial q}(\frac{dY}{ds}1,u_0) = C$$

Now recall that C is invertible, so that locally, $q \to \pi\psi_1(\frac{dY}{ds}1 + q, u_0)$ is a diffeomorphism.

More generally take $m \geq 2$, $\delta > 0$, and assume σ is a C^∞ mapping from $T^*_{y_0} M \times N$ into N such that

$$d^m(\psi_1(\frac{dY}{ds}1 + .,.), \sigma(.,.)) \leq \delta$$

Using the invertibility of C and the implicit function theorem, it is clear that if δ is small enough, there is $\delta' > 0, \eta > 0$ such that if $y \in M$ is such

$$d(y,y_0) < \delta'$$

the equation

$$\pi\sigma(q,u_0) = y \quad ; \quad |q| \leq \eta$$

has one unique solution which is a C^∞ function of y; moreover for $|q| \leq \eta$, the linear mapping

$$(4.46) \quad q' \in T^*_{y_0} M \to \pi^* \frac{\partial \sigma}{\partial q}(q,u_0)q'$$

has a uniformly bounded inverse.

$m \geq 2$, δ', η are now fixed. Assume that w^1 is such that

(4.47) $\quad \sup_{\substack{0 \leq s \leq 1 \\ 0 \leq t' \leq t}} d^m(\psi_s^{t'}(\sqrt{t'}dw^1, \frac{d\gamma}{ds}1+.,.), \psi_s(\frac{d\gamma}{ds}1+q,.)) \leq \delta$

Then if y is such that

$d(y,y_0) < \delta'$

for every $t' \leq t$, the equation

(4.48) $\quad \pi \psi_1^{t'}(\sqrt{t'}\,dw^1, \frac{d\gamma}{ds}1+q, u_0) = y \; ; \; |q| \leq \eta$

has one single solution.

<u>Definition 4.12</u> : $q(t',dw^1, y)$ denotes the unique solution of (4.48). $\hat{v}(t',dw^1,y)$ is the element of H_2

(4.49) $\quad v^2(t', dw^1, y) = \rho(q(t', dw^1, y))$

Of course, $q(t', dw^1, y)$ will be a C^∞ function of $(\sqrt{t'}, y)$.

f) <u>An asymptotic expression for $p_t(x_0,y_0)$</u>

Before giving the first global asymptotic expression for $p_t(x_0,y_0)$, we still need to prove an elementary lemma.

$g(x)$ is a ≥ 0 element of $C_b^\infty(R)$, which is 1 for $|x| \leq \frac{\delta}{2} \wedge \frac{\eta}{2}$, and 0 for $|x| \geq \delta \wedge \eta$. Also we assume that g decreases on R^+.

Set

(4.50) $\quad q(dw) = C^{-1} J_1(dw)$

Definition 4.13 : $G(t, dw^1)$ is the random variable

(4.51) $\quad G(t, dw^1) = g[\sup_{\substack{0 \leq s \leq 1 \\ t' \leq t}} d^m [\psi_s^{t'}(\sqrt{t'} \, dw^1, \frac{d\gamma}{ds}1 + ., .), \psi_s(\frac{d\gamma}{ds}1 + ., .)]]$

If $G(t, dw^1) \neq 0$, it is clear that (4.47) holds.

We now have :

Proposition 4.14 : The function $y \to E^{P_{u_0, y}^t} G(t, dw^1) g(|q(\sqrt{t} dw) - \frac{d\gamma}{ds}1|)$ is continuous on M.

Proof : For $u < 1$, set

$$(\dot{w}^{2,u})_s^i = <C^{-1} J_u(dw), h_s^i>$$

$$w_s^{1,u} = w_s - \int_0^s \dot{w}_h^{2,u} \, dh$$

$$q^u(\sqrt{t} \, dw) = C^{-1} J_u(\sqrt{t} \, dw)$$

$$G^u(t,dw) = g[\sup_{\substack{0 \le s \le u \\ t' \le t}} d^m [\psi_s^{t'}(\sqrt{t'} \, dw^{1,u}, \tfrac{d\gamma}{ds}1 + .,.), \psi_s(\tfrac{d\gamma}{ds}1+.,.)]]$$

Now $G^u(t,dw)$ is F_u-measurable, so that

(4.52) $\quad E^{\bar{P}_{u_0,y}^t}[G^u(t, dw) \quad g(|q^u(\sqrt{t} \, dw) - \tfrac{d\gamma}{ds}1|)] = E^P[G^u(t,dw) \, g(|q^u(\sqrt{t} \, dw)$

$$- \tfrac{d\gamma}{ds}1|) \, \frac{P_{t(1-u)}(x_u^t,y)}{P_t(x_0,y)}]$$

where of course $x_s^t = \pi \psi_s^t(\sqrt{t} \, dw, u_0)$. (4.52) is trivially a continuous function of y.

Using the arguments of the proof of Theorem 4.11 and time reversal it is not difficult to see that as u ↑↑1, the l.h.s. of (4.52) converges uniformly to the function considered in the Proposition. □

Recall that ρ has been defined in Definition 4.2.

We now define:

<u>Definition 4.15</u> : $\frac{\partial \psi_1^t}{\partial v}(\sqrt{t} \, dw^1, q, u_0)$ is the linear mapping from H in

$T \psi_1^t(\sqrt{t} \, dw^1, q, u_0)$

(4.53) $\quad v \to \psi_1^{t*} \int_0^1 (\psi_s^{t*-1}(\sqrt{t} \, dw^1, q, .) Y_i)(u_0) \, v^i \, ds$

$\widetilde{\pi^*\left(\frac{\partial \psi_1^t}{\partial v}\right)}(\sqrt{t} \, dw^1, q, u_0)$ is the mapping

$p \in T^*_{\pi\psi_1}(\sqrt{t} \, dw^1, q, u_0) M \to (<p, \pi^* \psi_1^{t*}(\psi_s^{t*-1}(\sqrt{t}dw^1,q,.)Y_i)(u_0) >) \in H$

If $\pi\psi_1^t(\sqrt{t}\ dw^1, \frac{dy}{ds}1 + q, u_0) = y_0$, $C(t, dw^1, q)$ is the linear mapping from $T_{y_0}^* M$ in $T_{y_0} M$

(4.54) $\quad p \in T_{y_0}^* M \to \pi^* \frac{\partial \psi_1^t}{\partial v}(\sqrt{t}\ dw^1, \frac{dy}{ds}1 + q, u_0)\ p(p)$

It is easy to check that if $\pi\psi_1^t(\sqrt{t}\ dw^1, \frac{dy}{ds}1 + q, u_0) = y_0$,

(4.55) $\quad C(t, dw^1, q) = \pi^* \frac{\partial \psi_1^t}{\partial q}(\sqrt{t}\ dw^1, \frac{dY}{ds}1 + q, u_0)$

Of course

$$C(0, dw^1, 0) = C$$

Moreover, by identifying $T_{y_0} M$ and $T_{y_0}^* M$, $C(t, dw^1, q)$ has a well-defined determinant which we write $\det C(t, dw^1, q)$.

We will note.

(4.56) $\quad p_t(x_0, y_0) \equiv A(t)$

if for any $k \in \mathbb{N}$, as $t \downarrow 0$

$$p_t(x_0, y_0) - A(t) = p_t(x_0, y_0)\ o(t^k)$$

We now have the following key result (which is the analogue of (0.21)).

Theorem 4.16 : As $t \downarrow 0$

$$(4.57) \quad p_t(x_0,y_0) \equiv \frac{[\det C]^{1/2}}{(\sqrt{2\pi t})^n} \int_\Omega \frac{\exp\{-\int_0^1 \frac{|\lambda+v_s^2(t,dw^1,y_0)|^2}{2t} ds\}}{\det C(t, dw^1, q(t,dw^1,y_0))} G(t,dw^1)$$

$$g(|q(t, dw^1, y_0)|) dP_1(w^1).$$

Proof : First note that the r.h.s. of (4.57) is well-defined. Indeed, if $G(t, dw^1) \neq 0$, by using the argument after (4.45), $q(t, dw^1, y_0)$ and $v^2(t,dw^1, y_0)$ are well-defined. Moreover from (4.46) and (4.54), $[\det C(t,dw^1,q(t,dw^1,y_0))]^{-1}$ is uniformly bounded.

We now prove (4.57). Take $f \in C_b^\infty(M)$ whose support is contained in $\{y; d(y_0,y) < \delta'\}$. Observe that in (4.50)

$$(4.58) \quad q(dw) = q(\dot{w}^2 ds)$$

Clearly

$$(4.59) \quad \int_\Omega (f(\pi\psi_1^t(\sqrt{t}\, dw, u_0))\, G(t,dw^1)\, g(|q(\sqrt{t}\, dw) - \tfrac{d\gamma}{ds}1|)\, dP(w)$$

$$= \int_M f(y)\, E^{\bar{P}_{u_0,y}^t}\, G(t,dw^1)\, g(|q(\sqrt{t}\, dw) - \tfrac{d\gamma}{ds}1|)\, p_t(x_0,y)\, dy$$

Since λ is in H_2, we have

(4.60) $$\int_0^1 <\lambda, \delta w> = \int_0^1 <\lambda, \dot{w}^2> ds.$$

Observe that w^1 is not modified when w_s is changed into $w_s + \frac{\lambda s}{\sqrt{t}}$, because $\lambda \in H_2$. Using (4.58), (4.60) and Girsanov's transformation, we see that (4.59) is also equal to

(4.61) $$\int_\Omega f(\pi \psi_1^t(\sqrt{t}\, dw, \frac{d\gamma}{ds}1, u_0))\, G(t, dw^1)$$

$$g(|q(\sqrt{t}(\dot{w}^2 ds))|) \exp\{-\int_0^1 [<\frac{\lambda, \dot{w}^2}{\sqrt{t}}> + \frac{1}{2}\frac{|\lambda|^2}{t}]\, ds\}\, dP(w)$$

We now split the measure P into the tensor product $P_1 \otimes P_2$, so that (4.61) is also equal to

(4.62) $$\int_{\Omega \times H_2} f(\pi \psi_1^t(\sqrt{t}\, dw^1 + \sqrt{t}v^2 ds, \frac{d\gamma}{ds}1, u_0))\, G(t, dw^1)$$

$$g(|q(\sqrt{t}v^2 ds)|) \exp\{-\int_0^1 \frac{|\lambda + \sqrt{t}\, v^2|^2 ds}{2t}\} \frac{dv^2\, dP_1(w^1)}{(\sqrt{2\pi})^n}$$

Obviously

(4.63) $$\psi_1^t(\sqrt{t}\, dw^1 + \sqrt{t}v^2 ds, \frac{d\gamma}{ds}1, u_0) = \psi_1^t(\sqrt{t}dw^1, q(\sqrt{t}v^2 ds) + \frac{d\gamma}{ds}1, u_0)$$

Consider the change of variables

(4.64) $$v^2 \to \pi \psi_1^t(\sqrt{t}\, dw^1, q(\sqrt{t}v^2 ds) + \frac{d\gamma}{ds}1, u_0) = y.$$

If $G(t,dw^1) \neq 0$, the statements after (4.45) hold. Moreover $f(y) \neq 0$ only if $d(y,y_0) < \delta'$. Finally $g(|q(\sqrt{t}\, v^2 ds)|) \neq 0$ only if $|q(\sqrt{t}\, v^2 ds)| \leq \eta$. It is obvious from Definition 4.12 that if the integrand in (4.62) is $\neq 0$, then if $\pi\psi_1^t(\sqrt{t}\, dw^1, q(\sqrt{t}\, v^2 ds)) + \frac{d\gamma}{ds}1, u_0) = y$

(4.65) $\quad q(\sqrt{t}\, v^2\, ds) = q(t, dw^1, y)$

$\quad\quad\quad \sqrt{t}\, v^2 = v^2(t, dw^1, y)$.

and (4.64) has a non 0 Jacobian.

Using the usual finite dimensional formula of change of variables, we see that (4.62) is equal to

(4.66) $\quad \dfrac{1}{(\sqrt{2\pi t})^n} \displaystyle\int_{\Omega \times M} \dfrac{f(y)\, G(t,dw^1) g(|q(t,dw^1,y)|)\, \exp\{-\int_0^1 \dfrac{|\lambda + v^2(t,dw^1,y)|^2}{2t} ds\}\, d\,P_1(w^1)\, dy}{\det [\pi^* \dfrac{\partial \psi_1^t}{\partial v}(\sqrt{t}dw^1, q(t,dw^1,y) + \dfrac{d\gamma}{ds}1,\, y)\, |_{H_2}]}$

In (4.66) $[\det \pi^* \dfrac{\partial \psi_1^t}{\partial v}(...)|_{H_2}]$ is the determinant of the corresponding linear mapping restricted to H_2.

Now it is trivial to check that this determinant is also equal to

(4.67) $\quad \dfrac{\det\, C(t,dw^1,q(t,dw^1,y))}{[\det\, C]^{1/2}}$

If we identify (4.59) and (4.66), and if we use (4.67), we find that

(4.68) $E^{\bar{P}^t_{u_0,y}}[G(t,dw^1) g(|q(\sqrt{t}\, dw) - \frac{d\gamma}{ds}1|)] p_t(x_0,y) =$

$$\frac{[\det C]^{1/2}}{(\sqrt{2\pi t})^n} \int \exp\{-\int_0^1 \frac{|\lambda + v^2(t,dw^1,y)|^2}{2t} ds\} \frac{G(t,dw^1) g(|q(t,dw^1,y)|)}{\det C(t,dw^1,q(t,dw^1,y^1))} dP_*(w^1)$$

a.e. on $\{y; d(y,y_0) < \delta'\}$

By Proposition 4.14 the l.h.s. of (4.68) is continuous in y. The r.h.s. being trivially continuous in y, there is equality everywhere in (4.68), in particular at $y = y_0$.

Now from Theorem 4.11, it immediately follows that for any $k \in \mathbb{N}$, as $t \downarrow\downarrow 0$

(4.69) $|E^{\bar{P}_{u_0,y_0}}[G(t,dw^1) g(|q(\sqrt{t}\, dw) - \frac{d\gamma}{ds}1|)] - 1| = o(t^k)$

Using (4.68) at y_0 and (4.69), (4.57) is proved. □

A useful by-product of the proof of Theorem 4.16 is as follows.

<u>Corollary</u> : Let K be a bounded measurable function on Ω . Then

(4.70) $p_t(x_0,y_0) E^{\bar{P}^t_{u_0,y_0}}[G(t, dw^1) g(|q\sqrt{t}\, dw) - \frac{d\gamma}{ds}1|) K(\sqrt{t}\, dw)]$

$$= \frac{[\det C]^{1/2}}{(\sqrt{2\pi t})^n} \int_\Omega \frac{\exp\{-\int_0^1 \frac{|\lambda+v^2(t,dw^1,y_0)|^2}{2t} ds\}}{\det C(t, dw^1, q(t,dw^1,y_0))}$$

$G(t,dw^1) g(|q(t,dw^1,y_0)|) K(\sqrt{t}\, dw^1 + (\lambda + v^2(t, w^1, y_0) ds) dP_1(w^1)$

Proof : Assume first that K is F_u-measurable with u < 1. Then the proof is identical to the proof of Theorem 4.16. Since both sides of (4.70) define positive measures, (4.70) immediately extends to a general K. □

Remark 6 : Of course G, g serve only as mollifiers and will later disappear (i.e. will be made equal to 1) when taking the Taylor expansion of $p_t(x_0, y_0)$. In particular, recall that g depends on δ, η. However we may take δ, η as small as we want. The possibility of choosing δ, η will be very useful when taking the Taylor expansion of $p_t(x_0, y_0)$. Also using Theorem 4.8 and Theorem I.2.1 in [10], it is very easy to see that for any $k \in N$, as $t \downarrow\downarrow 0$

(4.71) $P_1 [(G(t, dw^1) \neq 1) \cup (g(|q(t, dw^1, y_0)|) \neq 1)] = o(t^k)$

Of course the proof of (4.71) is much easier than the proof of (4.28) in Theorem 4.11, although the two results are obviously related.

(4.70) clearly shows that at least locally, $(w^1, v^2(t, dw^1, y_0))$ is a good parametrization of $\{w; \pi\psi_1^t(\sqrt{t}\, dw, u_0) = y_0\}$. In the sequel we will exploit only a small part of (4.70) which gives a precise description of the local conditional law $\bar{P}^t_{u_0, y_0}$.

Remark 7 : The same procedure would give a similar result for the semi-group p_t associated to the Riemann-Kodaira operator □ acting on forms.

Remark 8 : The representation (4.57) of $p_t(x_0, y_0)$ has all the nice tensor product properties which are exploited in Mc Kean-Singer [49].

Finally it must be pointed out that (4.70) can be expressed in terms of $\beta_{.} = u_0 w_{.}$ so that (4.70) will become a purely covariant expression computed along the geodesic γ and not depending on a peculiar choice of u_0.

g) The Jacobian of the exponential mapping in terms of a Brownian Bridge

Before expanding the r.h.s. of (4.57), we need an intermediary result, which is a natural extension of De Witt-Morette [19] - [20], De Witt-Morette, Maheshwari, Nelson [21], Elworthy-Truman [29].

Recall that x_0 and y_0 are assumed to be non conjugate.

Consider the equation of the Jacobi fields along γ

$$(4.72) \quad \frac{D^2 K}{Ds^2} + R(K, \frac{d\gamma}{ds}) \frac{d\gamma}{ds} = 0$$

$$K(0) = 0 \; ; \; \frac{DK}{Ds}(0) = e$$

(4.72) defines a linear map $e \to K_1 = K(e)$. K is a one to one mapping. det K is clearly equal to the Jacobian of the exponential mapping at $(x_0, \frac{d\gamma}{ds} 0)$.

We then have:

Theorem 4.17 : The following equality holds

$$(4.73) \quad [\det K]^{-\frac{1}{2}} = \int_\Omega \exp\{\frac{1}{2} \int_0^1 < \Omega(a,\lambda)\lambda, a > ds\} \; dQ(a)$$

There is $\mu > 1$ such that

$$(4.74) \quad \int_\Omega \exp\{\frac{\mu}{2} \int_0^1 < \Omega(a,\lambda)\lambda, a > ds\} \; dQ(a) < +\infty.$$

Proof : To prove (4.71) we use the stochastic calculus. Since $a'_s = a_{1-s}$ is a Brownian bridge, there is a n-dimensional Brownian motion B such that

$$da' = \frac{-a'}{1-s} ds + dB$$

$$a'(0) = 0$$

If $K'_s = u^{-1}_{1-s} K_{1-s} u_0$, we claim that $\frac{dK'_s}{ds} K'^{-1}_s$ is symmetric. We only need to prove that if $X, Y \in R^n$

(4.75) $\qquad < \frac{dK'_s}{ds} X, K'_s Y > = < K'_s X, \frac{dK'_s}{ds} Y > .$

Now (4.75) holds at s = 1. Moreover using (4.72), we see that both sides of (4.75) have the same derivatives. $\frac{dK'_s}{ds} K'^{-1}_s$ is then symmetric.

Using the stochastic calculus, it is easy to verify that if

$$M_s = \exp\{ \int_0^s \frac{1}{2} < \Omega(a'_h, \lambda)\lambda, a'_h > dh + \frac{1}{2} < (\frac{dK'_s}{ds} K'^{-1}_s + \frac{I}{1-s}) a'_s, a'_s > - \int_0^s \frac{Tr(\frac{dK'}{ds} k'^{-1} + \frac{I}{1-h})}{2} dh \}$$

M_s is a local martingale, and moreover

$$dM = M < (\frac{dK'}{ds} K'^{-1} + \frac{I}{1-s}) a'_s, \delta B_s >$$

$$M(0) = 1$$

Since $K'(1) = 0$, $\frac{dK'}{ds}1 = -I$, $\frac{dK'}{ds}s\ K_s'^{-1} + \frac{I}{1-s}$ is uniformly bounded. Much as in the proof of Theorem 4.10, and using the properties of the Girsanov transformation, one can easily prove that M is uniformly integrable. Clearly

$$(4.76) \quad \frac{1}{[\det K_0']^{1/2}} = \exp\{\frac{1}{2} \int_0^1 \text{Tr}\ [\frac{dK'}{ds}K'^{-1} + \frac{I}{1-h}]\ dh\}.$$

Since $E^Q M_1 = 1$ and $a_1' = 0$, (4.73) is a consequence of (4.76).

If $\mu > 1$, and if $\mu-1$ is small enough, x_0 and y_0 will still be non conjugate for the Sturm-Liouville operator $\frac{D^2}{Ds^2} + \mu R(.,\frac{d\gamma}{ds})\frac{d\gamma}{ds}$, so that (4.74) will also be finite. □

Remark 9 : Let z be a one dimensional Bes(3) process with $z(0) = 0$. Let T_1 be the first time at which z hits 1. Consider the differential equation

$$\frac{DH}{Ds} = \frac{1}{2} HR(., \frac{d\gamma}{ds}\ z_s)\ \frac{d\gamma}{ds}z_s$$

$$H(0) = I$$

$\frac{H_s K_{z_s}}{z_s}$ is easily seen to be a martingale, so that if $E\ [\sup_{0 \le s \le T_1} |H_s|] < +\infty$,

$$(4.77) \quad K_1^{-1} = E\ [H_{T_1}]$$

In the 1-dimensional case, the connection between (4.73) and (4.77) follows from the fact that the local time process $L_{T_1}(a)$ of z is a $\text{Bes}^2(2)$ bridge, with $L_{T_1}(0) = L_{T_1}(1) = 0$. In the general case, there is a "miracle" in (4.73) with no satisfactory probabilistic interpretation in connection with (4.77).

h) A path-integral proof of a result of Molchanov

Before expanding (4.57), we still must obtain the critical equivalent of $p_t(x_0,y_0)$ at $t \downarrow 0$, i.e. we must reobtain the result of Molchanov [54] (for a very simple proof, see Elworthy-Truman [30]).

The proof of this result will require a considerable effort, since we must prove very precise estimates on the expression (4.57). However, these estimates will later be essential to obtain the Taylor expansion of $p_t(x_0,y_0)$. Let us again insist on the fact that the methods of Schilder [57] are not applicable since none of the differentiability conditions of [57] are verified.

In the sequel, ∂^ℓ will denote the differential operator $\dfrac{\partial^\ell}{\partial(\sqrt{t})^\ell}$.

Theorem 4.18 : As $t \downarrow 0$

$$(4.78) \quad p_t(x_0,y_0) \sim \frac{1}{(\sqrt{2\pi t})^n [\det K]^{1/2}} \exp\{ \int_0^1 <b(\gamma_s), d\gamma_s> - \frac{d^2(x_0,y_0)}{2t} \}$$

Proof : In Definition 4.13, we will assume that $m \geq 3$. In the r.h.s. of (4.57), we only integrate in the region where $G(t,dw^1) \neq 0$, so that (4.47) is verified. Of course since the actual values of δ, η are irrelevant in the final result, we will choose δ, η as small as necessary.

We will write $v^2(t')$, $q(t')$, $C(t')$ instead of $v^2(t',dw^1,y_0)$, $q(t',dw^1,y_0)$, $C(t',dw^1,q(t',dw^1,y_0))$. Clearly

(4.79) $\quad \pi \psi_1^{t'} (\sqrt{t'}\, dw^1, q(t') + \frac{d\gamma}{ds}1, u_0) = y_0$

so that

(4.80) $\quad \pi^* \partial \psi_1^{t'} (\sqrt{t'}\, dw^1, q(t') + \frac{d\gamma}{ds}1, u_0) = 0$

At $t' = 0$, (4.80) writes

(4.81) $\quad \pi^* f_1^* \int_0^1 (f_s^{*-1} Y_i)(u_0)(dw^{1,i} + \partial v^{2,i}\, ds) = 0$

Since

(4.82) $\quad \pi^* f_1^* \int_0^1 (f_s^{*-1} Y_i)(u_0)\, dw^{1,i} = 0$

it is clear that $\partial v^2(0) = 0$.

Using Taylor's formula, there exists h such that $0 \leq h \leq t$ (which depends on w^1) and moreover

(4.83) $\quad \int_0^1 \frac{|\lambda + v^2(t)|^2}{2t}\, ds = \int_0^1 \frac{|\lambda|^2}{2t}\, ds + \frac{1}{2} \int_0^1 (\langle \lambda + v^2(h), \partial^2 v^2(h) \rangle + |\partial v^2(h)|^2)\, ds$

We will compute

(4.84) $\quad \frac{1}{2} \int_0^1 \langle \lambda, \partial^2 v^2(0) \rangle$

Set

(4.85) $\theta_{\cdot}^{(1)} = \theta(\partial \psi_{\cdot}^{t}(\sqrt{t}\ dw^1, q(t)+\frac{d\gamma}{ds}1, u_0)), \omega_{\cdot}^{(1)} = \omega(\partial \psi_{\cdot}^{t}(\sqrt{t}\ dw^1, q(t)+\frac{d\gamma}{ds}1, u_0))$

Using (2.2) we have

(4.86) $d\theta^{(1)} = (t\ \overline{\nabla b}\ \theta^{(1)} + 2\sqrt{t}\ \theta(b') + \partial\ v^2)\ ds + dw^1 + \omega^{(1)}((\lambda+v^2)ds$
$$+ \sqrt{t}\ dw^1)$$

$\theta^{(1)}(0) = 0$

$d\omega^{(1)} = \Omega((t\theta(b')+\lambda+v^2)\ ds + \sqrt{t}\ dw^1,\ \theta^{(1)})$

$\omega^{(1)}(0) = 0$

Of course $\theta^{(1)}(1) = 0$. By differentiating (4.86) again and using the fact that at $t = 0$ $\partial v^2 = 0$, we get

(4.87) $d\theta^{(2)} = (2\theta(b') + \partial^2 v^2)\ ds + \omega^{(2)}\lambda\ ds + \omega^{(1)}\ dw^1$

$\theta^{(2)}(0) = 0$

Since $\theta^{(1)}(1) = 0$, $\theta^{(2)}(1) = 0$. At $t = 0$, we see that

(4.88) $\frac{1}{2}\int_0^1 <\lambda,\ \partial^2 v^2>ds = \frac{1}{2}\int_0^1 <\lambda,\ d\theta^{(2)} - 2\theta(b')ds - \omega^{(2)}\lambda\ ds - \omega^{(1)}\ dw^1>$

Using Theorem 4.10, we now construct w^1 as in (4.23). Using the second line in (4.86), the antisymmetry of $\omega^{(2)}$, and integrating by parts in (4.88), we see that since at $t = 0$, $\theta^{(1)} = a$, $\omega^{(1)} = \alpha$, then

$$(4.89) \quad -\int_0^1 <\lambda, \theta(\dot{b})> ds - \frac{1}{2} \int_0^1 <\lambda, \omega^{(1)}(da - a\lambda\, ds)>$$

$$= -\int_0^1 <b(\gamma_s), d\gamma_s> + \frac{1}{2} \int_0^1 (<\Omega(\lambda,a)a, \lambda> - |a\lambda|^2) ds.$$

Now as $t \downarrow\downarrow 0$

$$(4.90) \quad \frac{G(t, dw^1)\, g(q(t))}{\det C(t)} \to \frac{1}{\det C} \quad \text{boundedly.}$$

Using (4.23) and (4.86), it is very easy to check that

$$(4.91) \quad K' = (\det C)^{-1/2}.$$

For a.e. w^1 such that as $t \downarrow\downarrow 0$, $G(t, dw^1)\, q(t) \to 1$, it is then clear that as $t \downarrow\downarrow 0$

$$(4.92) \quad \frac{1}{2} \int_0^1 (<\lambda + v^2(h), \partial^2 v^2(h)> + |\partial v^2(h)|^2)\, ds \to \frac{1}{2} \int_0^1 <\lambda, \partial^2 v^2(0)>.$$

Using Theorem 4.10, (4.83), (4.92), if it is feasible to take the limit under the expectation sign, we see that

(4.93) $(\det C)^{1/2} \int \exp\{-\frac{1}{2}\int_0^1 (<\lambda+v^2(h), \partial^2 v^2(h)> + |\partial v^2(h)|^2 \, ds\} \, G(t, dw^1)$

$\dfrac{g(|q(t)|)}{\det C(t)} \, dP_1(w^1) \to \int \exp \frac{1}{2} \int_0^1 < \Omega(\lambda, a)a, \lambda > ds \, dQ(a) \, \exp \int_0^1 <b(\gamma), d\gamma>$

By theorems 4.16, 4.17, we see that the Theorem will be proved if we justify (4.93).

We will show there exists $q > 1$ such that if δ, η are small enough

(4.94) $\int \exp\{-\frac{q}{2}\int_0^1 <\lambda+v^2(h), \partial^2 v^2(h)> ds\} \, G(t, dw^1) \, dP_1(w^1)$

is uniformly bounded as $t \downarrow\downarrow 0$. This immediately implies uniform integrability in (4.93), and so (4.93) is an obvious consequence of (4.94).

To control $\partial^2 v^2(h)$, we first need to control $\partial v^2(h)$, since $\partial^2 v^2(h)$ in fact depends on $\partial v^2(h)$. We will first control $\partial v^2(h)$, then $\partial^2 v^2(h) - \partial^2 v^2(0)$, and finally $\partial^2 v^2(0)$.

A) <u>Control of $\partial v^2(h)$</u>

Set

(4.95) $v_0 = f_1(u_0)$

Let f' be the flow of diffeomorphisms of N associated to the differential equation

(4.96) $dv = -Y_i(v) \lambda^i \, ds$

Clearly if $q \in T^*_{y_0} N$

(4.97) $\quad <Cq,q> = \int_0^1 <\pi^*(f'^{*-1}_s Y_i)(v_0), q>^2 \, ds$

Since C is positive definite, it is clear that there exists β such that $0 < \beta < 1$, and moreover the quadratic form C^β on $T^*_{y_0} M$ defined by

(4.98) $\quad <C^\beta q,q> = \int_0^1 <\pi^*(f'^{*-1}_s Y_i)(v_0), q>^2 \, ds$
$\qquad\qquad\qquad 1-\beta$

is also positive definite.

If $G(t,dw^1) \neq 0$, it is easy to see that for $t' \leq t$

(4.99) $\quad \partial \psi_1^{t'}(\sqrt{t'} \, dw^1, q(t') + \frac{d\gamma}{ds}1, u_0) =$

$\qquad \psi_1^{t'*} [\int_0^1 (\psi_s^{t'*-1} Y_i)(u_0)(dw^{1,i} + \partial v^{2,i} \, ds)$.

$\qquad + 2\sqrt{t'} \int_0^1 (\psi_s^{t'*-1} b')(u_0) \, ds]$

where $\psi_s^{t'*}$ is of course $\psi_s^{t'*}(\sqrt{t'} \, dw^1, q(t') + \frac{d\gamma}{ds}1, u_0)$.

It is essential to note that the r.h.s. of (4.99) is <u>not</u> a bona fide (i.e. non anticipating) stochastic integral, but is well defined either by the method of Proposition 3.9 and (4.32), or by taking the adequate regularization of certain standard stochastic integrals as in [10]-III.

In the sequel, C will denote uniform constants, which vary from place to place.

Clearly from the definition of P_1

$$(4.100) \quad \pi^* \psi_1^{t'*} \int_0^1 (\psi_s^{t'*-1} Y_i)(u_0) \, dw^{1,i} =$$

$$\pi^* \psi_1^{t'*} \int_0^1 ((\psi_s^{t'*-1} Y_i)(u_0) - (f_s^{*-1} Y_i)(u_0)) \, dw^{1,i} +$$

$$+ \pi^*(\psi_1^{t'*} - f_1^*) \int_0^1 (f_s^{*-1} Y_i)(u_0) \, dw^{1,i}$$

From (4.99), we see that if $G(t, dw^1) \neq 0$, for $t' \leq t$

$$(4.101) \quad |\partial v^2(t')| \leq C \, |\pi^* \psi_1^{t'*} \int_0^1 (\psi_s^{t'*-1} Y_i(u_0) - (f_s^{*-1} Y_i)(u_0)) \, dw^{1,i}| +$$

$$+ C \, |\pi^*(\psi_s^{t'*} - f_1^*) \int_0^1 (f_s^{*-1} Y_i)(u_0).dw^{1,i}| + C\sqrt{t}.$$

If $G(t, dw^1) \neq 0$, for $t' \leq t$

$$(4.102) \quad |\psi_1^{t'*} \int_0^\beta ((\psi_s^{t'*-1} Y_i)(u_0) - (f_s^{*-1} Y_i)(u_0)).dw^{1,i}|$$

$$\leq C \, | \int_0^\beta ((\psi_s^{*t'-1} Y_i)(u_0) - (f_s^{*-1} Y_i)(u_0)).dw^{1,i}|$$

Also it should be pointed out that using the results of [10], in (4.100)-(4.102), the Stratonovitch integrals are in fact Itô integrals so that we may replace everywhere $dw^{1,i}$ by $\delta w^{1,i}$.

For $q \in T^*_{y_0} M$, $t' \leq t$, set

(4.103) $K^{t'}_{i,s}(q) = (\psi^{t'*-1}_s (\sqrt{t}dw^1, q + \frac{d\gamma}{ds}1,.)Y_i)(u_0) - (f^{*-1}_s Y_i)(u_0)$.

$L^{t'}_h(q) = \int_0^h K^{t'}_{i,s}(q) \delta w^{1,i}$

Now $L^{t'}(q)$ is a bona fide stochastic integral. We will show that for any $p > 1$, if δ, η are small enough

(4.104) $\int \exp \{p \sup_{\substack{t' \leq t \\ |q| \leq \eta}} |L^{t'}_\beta(q)|^2\} G(t,dw^1) dP_1(w^1)$.

is uniformly bounded as $t \Downarrow 0$.

Set

(4.105) $G'(t,dw^1) = g (\sup_{\substack{0 \leq s \leq \beta \\ t' \leq t}} [d^m(\psi^{t'}_s (\sqrt{t} \cdot dw^1, \frac{d\gamma}{ds}1 +.,.), \psi_s(\frac{d\gamma}{ds}1+.,.))])$

Clearly (4.104) is dominated by

(4.106) $\int \exp \{p \sup_{\substack{t' \leq t \\ |q| \leq \eta}} |L^{t'}_\beta(q)|^2\} G'(t,dw^1) dP_1(w^1)$.

Let C_s be the quadratic form on $T^*_{y_0} M$ defined by

(4.107) $<C_s q,q> = \int_s^1 <\pi^*(f^*_1 f^{*-1}_h Y_i)(u_0), q>^2 dh$

Clearly for $s \leq \beta$, C_s has a uniformly bounded inverse. Moreover it is easy to see that on $B(w_s^1 | s \leq \beta)$, P_1 is absolutely continuous with respect to P, and that

$$(4.108) \quad \frac{dP_1}{dP} \mid B(w_s^1 | 0 \leq s \leq \beta) = \frac{(\det C)^{1/2}}{(\det C_\beta)^{1/2}} \exp \{ -\frac{1}{2} < C_\beta^{-1} \int_0^\beta \pi \tilde{f}_1^*(f_s^{*-1} Y_i)$$

$$(u_0) \cdot dw^{1,i}, \int_0^\beta \pi^* f_1^*(f_s^{*-1} Y_j)(u_0) \cdot dw^{1,j} > \}$$

and the r.h.s. of (4.108) is uniformly bounded.

Instead of (4.106), we may as well estimate

$$(4.109) \quad \int \exp \{ p \sup_{\substack{t' \leq t \\ |q| \leq \eta}} | L_\beta^{t'}(q) |^2 \} G'(t, dw^1) \, dP(w^1)$$

We first estimate for $t' \leq t$, $|q| \leq \eta$

$$(4.110) \quad \int \exp p \, | L_\beta^{t'}(q) |^2 \, G'(t, dw^1) \, dP(w^1).$$

Now

$$(4.111) \quad [L_\beta^{t'}(q)]^2 = 2 \int_0^\beta L_s^{t'}(q) \, K_{i,s}^{t'}(q) \, \delta w^{1,i} + \int_0^\beta (K_{i,s}^{t'}(q))^2 \, ds.$$

If $G'(t,dw^1) \neq 0$, $\int_0^\beta (K_{i,s}^{t'}(q))^2 \, ds$ is uniformly bounded, and so we can neglect this term in the estimation process. To estimate (4.110) we only need to estimate for any $p > 1$

$$(4.112) \quad \int \exp \{p \int_0^\beta L_s^{t'}(q) \, K_{i,s}^{t'}(q) \, \delta w^{1,i}\} \, G'(t,dw^1) \, dP(w_1)$$

Now using Cauchy-Schwarz's inequality and the fact that $G' \leq 1$, (4.112) is dominated by

$$(4.113) \quad [\int \exp \{2p \int_0^\beta L_s^{t'}(q) \, K_{i,s}^{t'}(q) \, \delta w^{1,i} - 2p^2 \int_0^\beta (L_s^{t'}(q) K_{i,s}^{t'}(q))^2 ds\}$$

$$dP(w^1)]^{1/2} \, [\int \exp \{2p^2 \int_0^\beta (L_s^{t'}(q) \, K_{i,s}^{t'}(q))^2 \, ds\} \, G'(t,dw^1) \, dP(w^1)]^{1/2}$$

Now in the first term $[..]$ in (4.113) we exactly find a Girsanov exponential (with respect to the Brownian motion w^1), so that since Girsanov exponentials are supermartingales, the first term $[...]$ is ≤ 1.

Moreover if $G'(t,dw^1) \neq 0$, $|K_{i,s}^{t'}(q)| \leq C(\delta+\eta)$. To dominate (4.112), we must estimate

$$(4.114) \quad \int \exp\{p \, (\delta+\eta)^2 \int_0^\beta (L_s^{t'}(q))^2 \, ds\} \, G'(t,dw^1) \, dP(w^1)$$

(of course p varies from place to place).

Using Jensen's inequality, we see that (4.114) is dominated by

$$(4.115) \quad \frac{1}{\beta} \int_0^\beta ds \int \exp\{p(\delta+\eta)^2 \beta (L_s^{t'}(q))^2\} G'(t,dw^1) \, dP(w^1).$$

Now a classical inequality on Itô's integrals shows that if $s \leq \beta$

$$(4.116) \quad P[\,|L_s^{t'}(q)| \geq a \,;\, G'(t,dw_1) \neq 0\,] \leq 2 \exp\left[-\frac{a^2}{2m\, c^2 (\delta+\eta)^2}\right]$$

From (4.116) it is then obvious that if δ, η are small enough, (4.115) will be uniformly bounded. We have then found that, for any $p > 1$, if δ, η are small enough, (4.110) is uniformly bounded.

By using the results of [10]-Chapter I and II, it is also very easy to prove that for $t', t'' \leq 1$, q, q' such that $|q'| \leq \eta$, $|q''| \leq \eta$, $p > 1$

$$(4.117) \quad \int |(L_\beta^{t'}(q'))^2 - (L_\beta^{t''}(q''))^2|^p \, dP(w^1) \leq C_p \, [|\sqrt{t'} - \sqrt{t''}|^p + |q' - q''|^p].$$

Now

$$(4.118) \quad \int |\exp(L_\beta^{t'}(q'))^2 - \exp(L_\beta^{t''}(q''))^2|^p \, G'(t,dw^1) \, dP(w^1)$$

$$\leq \int |\exp(L_\beta^{t'}(q'))^2 - \exp(L_\beta^{t''}(q''))^2|^p \, 1_{|L_\beta^{t'}(q') - L_\beta^{t''}(q'')| \geq 1}$$

$$G'(t,dw^1) \, dP(w^1) + C \int \exp\{p(L_\beta^{t'}(q'))^2\} \, |(L_\beta^{t'}(q'))^2 - (L_\beta^{t''}(q''))^2|^p \, G'(t,dw^1) \, dP(w^1)$$

Using the uniform bounds on (4.110), (4.117), Tchebytchev's and Hölder's inequalities, we find that for δ, η small enough (depending on p), (4.118) is dominated by

(4.119) $\quad C \left[|\sqrt{t'} - \sqrt{t''}|^p + |q' - q''|^p \right]$

Using Garsia-Rodemich-Rumsey's theorem [82] (also see [64]), we find from (4.118),(4.119) that for any $p > 1$, for δ, η small enough

(4.120) $\displaystyle\sup_{\substack{0 \le t' \le t'' \le t \\ |q'|, |q''| \le \eta}} \frac{|\exp(L_\beta^{t'}(q'))^2 - \exp(L_\beta^{t''}(q''))^2|}{[|\sqrt{t'} - \sqrt{t''}| + |q' - q''|]^{1/2}} G'(t, dw^1)$

is in $L_p(P)$ (with a norm which can be bounded independently of t). Now recall that

(4.121) $\quad L_\beta^0(0) = 0$

From (4.120), (4.121), we find that if δ, η are small enough, (4.109) and (4.106) are uniformly bounded. The statement before (4.104) is proved.

In the first term of the r.h.s. of (4.101), we have controlled $\int_0^\beta \ldots$. We now must control $\int_\beta^1 \ldots$ We will use time reversal.

Set

(4.122) $\quad u^{t'} = \psi_1^{t'}(\sqrt{t'} dw^1, q(t') + \frac{d\gamma}{ds} 1, u_0).$

$\quad w_s^{'1} = w_1^1 - w_{1-s}^1$

Clearly if $\beta' = 1-\beta$

$$(4.123) \quad \pi^* \psi_1^{t'*} \int_\beta^1 ((\psi_s^{t'*-1} Y_{\cdot i})(u_0) - (f_s^{*-1} Y_{\cdot i})(u_0)) \, dw^{1,i} =$$

$$\pi^* \int_0^{\beta'} ((\psi_s^{t'*-1} Y_{\cdot i})(u^{t'}) - (f_s^{*-1} Y_{\cdot i})(u^{t'})) \, dw^{1,i} +$$

$$+ \pi^* \int_0^{\beta'} ((f_s^{'*-1} Y_{\cdot i})(u^{t'}) - (f_s^{'*-1} Y_{\cdot i})(v_0)) \cdot dw^{1,i} -$$

$$- \pi^* (\psi_1^{t'*} - f_1^*)(u_0) \int_0^{\beta'} (f_{1-s}^{*-1} Y_{\cdot i})(u_0) \cdot dw^{1,i}$$

Of course in (4.123), we can replace everywhere dw^1 by δw^1.

If the law of w^1 is P, the law of w'^1 is still P. Now P_1 is exactly P conditional on

$$(4.124) \quad \pi^* \int_0^1 (f_s^{'*-1} Y_{\cdot i})(v_0) \cdot dw^{1,i} = 0.$$

Recall that β has been choosen in such a way that C^β (defined in (4.98) is positive definite. Using (4.124), and the argument in (4.107)-(4.108), it is clear that on $B(w_s'^1 | 0 \leq s \leq \beta')$, P_1 and P are equivalent and that the corresponding density is uniformly bounded.

For $t' \leq t$, $q \in T_{y_0}^* M$, $u' \in N$ such that $\pi u' = y_0$, set

$$(4.125) \quad K_{i,s}^{'t}(q,u') = \pi^*[(\psi_s^{'t'*-1}(\sqrt{t'} \, dw^1, q+\tfrac{d\gamma}{ds}1, \cdot) Y_{\cdot i})(u') - (f_s^{'*-1} Y_{\cdot i})(u')]$$

$$L_s^{'t}(q,u') = \int_0^s K_{i,s}^{'}(q,u') \delta w^{1,i}$$

If $G(t,dw^1) \neq 0$, it is not hard to show that if $|q| \leq \eta$, $|u'-v_0| \leq \delta$,

$|K_{i,s}^{t'}(q,u')| \leq C(\delta+\eta)$.

Using now (t',q,u') instead of (t',q), we can restart on (4.125) the procedure used in (4.108)-(4.111) with the same conclusions.

A similar reasoning applies on the second term on the r.h.s. of (4.123). Finally if $G(t,dw^1) \neq 0$, using integration by parts, we see that

$$(4.126) \quad |\pi^*(\psi_1^{t'} - f_1^*)(u_0) \int_0^\beta f_{1-s}^{*-1} Y_i(u_0) \cdot dw'^{1,i}| \leq C \delta \sup_{0 \leq s \leq \beta'} |w_s'^{1,i}|.$$

For $\delta > 0$ small enough

$$(4.127) \quad \int \exp[p \, \delta^2 \sup_{0 \leq s \leq \beta'} |w_s'^{1,i}|^2] dP(w'^1) < +\infty.$$

A similar reasoning can be used on the second term in the r.h.s. of (4.101)

From Hölder's inequality, we then easily conclude that for any $p > 1$, if δ, η are small enough,

$$(4.128) \quad \int \exp\{\frac{p}{2} \int_0^1 |\partial v^2(h)|^2 \, ds\} \, G(t,dw^1) \, dP(w_1).$$

is uniformly bounded as $t \downarrow 0$.

B) Control of $<\lambda, \partial^2 v^2(h) - \partial^2 v^2(0)>$

We will now prove that for any $p > 1$, if δ, η are small enough,

(4.129) $\quad \int \exp\{\frac{p}{2} \mid \int_0^1 <\lambda, \partial^2 v^2(h) - \partial^2 v^2(0)> ds\mid\} \; G(t, dw^1) \; dP_1(w^1).$

is uniformly bounded.

To simplify the computations which follow, we will assume that $b = 0$,

To prove (4.129), we first need to know how $\partial^2 v^2(t')$ looks like. Using (4.80)-(4.99), we have

(4.130) $\quad \theta(\psi_1^{t'*} \int_0^1 (\psi_s^{t'*-1} Y_i)(u_0) (dw^{1,i} + \partial v^{2,i} ds)) = 0$

or equivalently

(4.131) $\quad \theta \; (\int_0^1 (\psi_s^{t'*-1} Y_i)(u^{t'}) (dw_s^{1,i} + \partial v_{1-s}^{2,i} ds)) = 0.$

When differentiating (4.131) we must take into account the variation of $\sqrt{t}'$ (keeping $u^{t'}$ fixed) and the variation of $u^{t'}$.

To compute the differential of (4.131), we will use the following notation. Namely if $Y(u)$ is a vector field on N, for every $u \in N$, $\overline{Y}^u(v)$ denotes the vector field defined by

$$\theta(\overline{Y}^u(v)) = 0, \quad \omega(\overline{Y}^u(v)) = \omega(Y(u))$$

In particular we can define the Lie Bracket $[\overline{Y}^u, Y]$ and its value at u $[\overline{Y}^u, Y](u)$.

By differentiating (4.131), we get

$$(4.132) \quad \theta \left\{ \int_{0 \le s \le s' \le 1} -[(\psi_s'^{t'*-1} Y_j)(u^{t'}), (\psi_{s'}^{t'*-1} Y_i)(u^{t'})] (dw_s'^{1,j} + \partial v_{1-s}^{2,j} ds) (dw_{s'}^{1,i} + \partial v_{1-s'}^{2,i} ds') \right\} +$$

$$\theta \left\{ [\int_0^1 (\psi_s'^{t'*-1} Y_i)(u^{t'}) (dw_s'^{1,i} + \partial v_{1-s}^{2,i} ds) \int_0^1 (\psi_s'^{t'*-1} Y_j)(u^{t'}) (dw_s'^{1,j} + \partial v_{1-s}^{2,j} ds)] \right\} +$$

$$+ \theta \left\{ \int_0^1 (\psi_s'^{t'*-1} Y_i)(u^{t'}) \partial^2 v_{1-s}^{2,i} ds \right\} = 0$$

and of course (4.132) determines $\partial^2 v^2(t')$. We claim that we can do on

$$(4.133) \quad \partial^2 v^2(t') - \partial^2 v^2(0)$$

the same sort of manipulations which we did on $\partial v^2(t')$, so that the desired result on (4.129) will hold.

Namely observe that we can replace dw'^1 by $\delta w'^1$ in (4.132) (because this is already true on (4.131)). Using (4.132) we can evaluate (4.133) in terms of stochastic integrals with respect to w^1, where the integrands are arbitrarily small when $G(t, dw^1) \ne 0$ (this by adequately choosing δ, η).

Now we can control all the terms of the type $\int_0^1 f \delta w'^1 \int_0^1 g \delta w'^1$ using the trivial

$$(4.134) \quad |\int_0^1 f \delta w'^1 \int_0^1 g \delta w'^1| \le \frac{1}{2}(|\int_0^1 f \delta w'^1|^2 + |\int_0^1 g \delta w'^1|^2)$$

(this is the case in particular when ∂v^2 appears in (4.132)).

In calculating (4.132) there are also terms of a type which we have not yet met i.e. terms like

$$(4.135) \quad \int_{0 \leq s \leq s' \leq 1} f_s \, g_{s'} \, \delta w_s^{'1} \, \delta w_{s'}^{'1}.$$

(4.135) is also equal to

$$(4.136) \quad \int_{0 \leq s \leq s' \leq \beta'} f_s \, g_{s'} \, \delta w_s^{'1} \, \delta w_{s'}^{'1} + \int_{\beta' \leq s \leq s' \leq 1} f_s \, g_{s'} \, \delta w_s^{'1} \, \delta w_{s'}^{'1}$$

$$+ \int_{0 \leq s \leq \beta'} f_s \, \delta w_s^{'1} \int_{\beta' \leq s' \leq 1} g_{s'} \, \delta w_{s'}^{'1}$$

Each of the first two terms in the r.h.s. of (4.136) is controllable using the techniques of (4.103)-(4.121). The third term is controlled using (4.134).

We then easily find that for any $p > 1$, there is δ, η such that (4.129) is uniformly bounded.

Also note that if $G(t, dw^1) \neq 0, |v_s^2(t)| \leq C\delta$. By using the techniques we used to prove (4.104), (4.129), it is easy to see that for any $p > 1$, if δ, η are small enough

$$(4.137) \quad \int \exp\{\tfrac{p}{2} | \int_0^1 <v^2(h), \partial^2 v^2(h)> ds |\} G(t, dw^1) dP_1(w^1)$$

is uniformly bounded.

c) **Control of $\partial^2 v^2(0)$**

Using (4.128), (4.129), (4.137) and Hölder's inequality, it is clear that in order to prove (4.94), we only need to show that for one $p' > 1$

$$(4.138) \quad \int \exp{-\tfrac{p'}{2} \int_0^1 <\lambda, \partial^2 v^2(0)> dP_1(w^1) < +\infty}.$$

Using (4.24), (4.89), we only need to show that for one $p'>1$

$$(4.139) \quad \int \exp\{(1 - \frac{p'}{2}) \int_0^1 <\Omega(\lambda,a)\,a,\lambda> ds + \frac{p'-1}{2} \int_0^1 |\alpha\lambda|^2 ds\} dQ(a) < +\infty.$$

Recall that $\mu > 1$ has been defined in Theorem 4.17. Using Hölder's inequality we see that if $p' < 2$, the l.h.s. of (4.139) is dominated by

$$(4.140) \quad \left[\int \exp\{\frac{\mu}{2} \int <\Omega(\lambda,a)a,\lambda> ds\} dQ(a) \right]^{\frac{2-p'}{\mu}}$$

$$\left[\int \exp\{\frac{\mu(p'-1)}{2(\mu-2+p')} \int_0^1 |\alpha\lambda|^2 ds\} dQ(a) \right]^{\frac{\mu-2+p'}{\mu}}$$

Now the first term in (4.140) is $< +\infty$ by Theorem 4.17. Moreover since $\mu > 1$, as $p' \downarrow\downarrow 1$, $\frac{\mu(p'-1)}{2(\mu-2+p')} \downarrow\downarrow 0$.

Finally from Theorem 4.10, it is clear that

$$\int_0^1 |\alpha\lambda|^2 \leq C \sup_{0 \leq s \leq 1} |a_s|^2$$

Now if $\xi > 0$ is small enough, it is trivial to see that

$$(4.141) \quad \int \exp\{\xi \sup_{0 \leq s \leq 1} |a_s|^2\} dQ(a) < +\infty$$

From (4.141), we see that for $p' > 1$ small enough (4.139) is $< +\infty$.
The Theorem is proved. □

i) Taylor expansion of $p_t(x_0, y_0)$

We first give a Definition.

__Definition 4.19__ : If $G(t, dw^1) \neq 0$, $E(t, dw^1)$ is defined by

$$(4.142) \quad E(t, dw^1) = \int_0^1 \frac{|\lambda + v^2(t)|^2}{2t} ds - \int_0^1 \frac{|\lambda|^2}{2t} ds - \frac{1}{2} \int_0^1 <\lambda, \partial^2 v^2(0)> ds.$$

If $G(t, dw^1) \neq 0$, $E(t, dw^1)$ has a Taylor expansion of the type

$$(4.143) \quad E(t, dw^1) = \sum_1^N E_{k+2}(dw^1) \, t^{k/2} + o(t^{N/2})$$

In (4.143), if $b = 0$, $E_k(dw^1)$ is a k-linear functional of dw^1, which is a sum of multiple iterated stochastic integrals

$$(4.144) \quad E_k(dw^1) = \sum_{j_1 \cdots j_k} \int_{0 \leq t_1 \cdots \leq t_k \leq 1} E_k^{j_1 \cdots j_k}(t_1, \ldots, t_k) dw_{t_1}^{1,j_1} \cdots dw_{t_k}^{k,j_k}$$

where the $E_k^{j_1 \cdots j_k}$ are (deterministic) bounded measurable functions. If $b \neq 0$, some of the dw_s^{1,j_ℓ} are replaced by components of $b(\gamma_s) ds$ or the covariant derivatives of b.

Similarly if $G(t, dw^1) \neq 0$, for any N, we have the Taylor expansion.

$$(4.145) \quad \det C(t, dw^1, q(t, dw^1, y_0)) = \det C + \sum_1^N c_k(dw^1) \, t^{k/2} + o(t^{N/2})$$

If $G(t, dw^1) \neq 0$, we have the Taylor expansion

$$(4.146) \quad \frac{\det C \exp[-E(t,dw^1)]}{\det C(t, dw^1, q(t,dw^1,y_0))} = 1 + \sum_1^N d_k(dw^1) \, t^{k/2} + o(t^{N/2})$$

When $b = 0$, the $d_k(dw^1)$ are also of the type (4.144), with the obvious modifications when $b \neq 0$.

<u>Proposition 4.20</u> : For any $k \in N$, $E_k(dw^1)$, $c_k(dw^1)$, $d_k(dw^1)$ are in all the $L_p(\Omega, P_1)$ ($1 \leq p < +\infty$). Similarly, if w^1 is defined on (Ω, Q) by (4.23), for any $k \in N$, $E_k(dw^1)$, $c_k(dw^1)$, $d_k(dw^1)$ are in all the $L_p(\Omega, Q)$ ($1 \leq p < +\infty$).

<u>Proof</u> : We may express w^1 in terms of w as in (4.19). If we enlarge the natural filtration of w so that it contains $J_1(dw)$, w is still a semi-martingale so that the stochastic integrals $E_k(dw^1), c_k(dw^1), d_k(dw^1)$ are still bona fide stochastic integrals for the expanded filtration. Noting that $J_1(dw)$ will factor every time that $\dot{w}^2$ appears, we finally obtain sums of products of standard stochastic integrals with respect to w, which are in all the $L_p(\Omega, P)$ ($1 \leq p < +\infty$) by Burkholder-Davis-Gundy's inequalities.

In (4.23), we may write $a_s = w_s - sw_1$. The same argument as before finishes the proof of the Proposition. □

We now have the key result of this section:

Theorem 4.21 : For any N, as $t \downarrow\downarrow 0$

$$(4.147) \quad p_t(x_0, y_0) = \frac{1}{(\sqrt{2\pi t})^n} \exp\{-\frac{d^2(x_0, y_0)}{2t} + \int_0^1 <\dot{t}(\gamma_s), d\gamma_s>\}$$

$$\left[\int \exp\{\frac{1}{2}\int_0^1 <\Omega(\lambda, a)a, \lambda>ds\}(1 + \sum_1^N d_{2k}(dw^1)t^k)dQ(a) + o(t^N)\right]$$

Proof : Observe that the r.h.s. of (4.147) makes good sense because of (4.74), Proposition 4.20 and Hölder's inequality.

We now prove (4.147). Recall that in Definition 4.13 $G(t, dw^1)$ depended on $m \in \mathbb{N}$. We take here $m \geq 4N$.

We now proceed formally as in Schilder [57].
Take $p' > 1$ such that (4.139) holds. Let $p > 1$ be such that $\frac{2}{p} + \frac{1}{p'} < 1$.

We will assume that δ, η have been chosen small enough so that the statements before (4.128), (4.129), (4.137) hold for the previous choice of p. Set

$$(4.148) \quad k(t, dw^1) = -\frac{E(t, dw^1)}{\sqrt{t}}$$

From Taylor's formula

$$(4.149) \quad e^{\sqrt{t} k(t, dw^1)} = 1 + t^{1/2}k(t, dw^1) + \ldots + \frac{t^N k^{2N}(t, dw^1)}{(2N)!} + R_{2N}(t, dw^1)$$

and moreover

(4.150) $\quad |R_{2N}(t,dw^1)| \leq |E(t, dw^1)|^{2N+1} e^{\sqrt{t}k^+(t,dw^1)}$

Now from (4.83), we know that if $G(t,dw^1) \neq 0$, for one (random) h such that $0 \leq h \leq t$

(4.151) $\quad E(t,dw^1) = \frac{1}{2} \int_0^1 (|\partial v^2(h)|^2 + <\lambda, \partial^2 v^2(h) - \partial^2 v^2(0)>$
$\qquad\qquad\qquad + <v^2(h), \partial^2 v^2(h)>)ds$

so that

(4.152) $\quad \sqrt{t}\, k^+(t,dw^1) \leq \frac{1}{2}[|\int_0^1 <\lambda, \partial^2 v^2(h) - \partial^2 v^2(0)> ds| + |\int_0^1 <v^2(h), \partial^2 v^2(h)> ds|]$

Using (4.129), (4.137)-(4.150), and Hölder's inequality, we find that if r is defined by

(4.153) $\quad \frac{2}{p} + \frac{1}{p'} + \frac{1}{r} = 1$

then

(4.154) $\quad \left| \int \frac{\exp - \frac{1}{2} \int_0^1 <\lambda, \partial^2 v^2(0)>ds \ \ G(t,dw^1) \ g(|q(t,dw^1)|)}{\det C(t, dw^1, q(t, dw^1, y_0))} \right.$

$\qquad\qquad \left. R_{2N}(t,dw^1) \ dP_1(w^1) \right| \leq C \left[\int |E(t, dw^1)|^{(2N+1)r} G(t,dw^1) \ dP_1(w^1) \right]^{1/r}$

Clearly if $G(t,dw^1) \neq 0$, for $t' \leq t$

(4.155) $\quad \partial^3 \int_0^1 \frac{|\lambda+v^2(t')|^2}{2} ds = \int_0^1 (3<\partial v^2(t'), \partial^2 v^2(t')> + <\lambda+v^2(t'), \partial^3 v^2(t')>)ds$

Moreover, we know that if $G(t,dw^1) \neq 0$, for one (random) h' such that $0 \leq h' \leq t$

$$(4.156) \quad E(t,dw^1) = \frac{t^{1/2}}{3!} \partial^3 \left[\int_0^1 \frac{|\lambda+v^2(h')|^2}{2} ds \right]$$

Using (4.155), inequalities like (4.117) and Garsia-Rodemich-Rumsey's theorem [81], like in (4.120), it is not difficult to prove that for any $p'' > 1$

$$(4.157) \quad \int \sup_{0 \leq t' \leq t} |\partial^3 \int_0^1 \frac{|\lambda+v^2(t')|^2}{2} ds |^{p''} G(t,dw^1) \, dP_1(w^1)$$

is uniformly bounded (as $t \downarrow 0$). Note that the proof does not involve careful estimates like (4.110) but is essentially based on the simple estimates of [10]-Chapter I and II and inequalities like (4.117).

Using (4.156), we see that (4.154) is dominated by $C \, t^{\frac{2N+1}{2}}$.

From Theorem 4.16, (4.71), (4.156)-(4.157), we see that

$$(4.158) \quad p_t(x_0,y_0) = \frac{(\det C)^{1/2}}{(\sqrt{2\pi t})^n} \exp\{\frac{-d^2(x_0,y_0)}{2t}\}$$

$$\int \frac{\exp\{-\frac{1}{2}\int_0^1 <\lambda, \partial^2 v^2(0)> ds\}}{\det C(t,dw^1, q(t,dw^1,y_0))} \left((1+t^{1/2} k(t,dw^1)+\ldots+ \frac{t^N k^{2N}(t,dw^1)}{(2N)!} \right.$$

$$\left. + o(t^N) \right) G(t,dw^1) \, dP_1(w^1)$$

Now if $G(t, dw^1) \neq 0$, we can take the Taylor expansion of $E(t,dw^1)$ like in (4.143), so that

$$(4.159) \quad k(t, dw^1) = - \sum_{0}^{2N} E_{k+3}(dw^1) \, t^{k/2} + t^{\frac{2N+1}{2}} S_N(t, dw^1)$$

Still proceeding as for (4.157), it can be proved that for any $p'' > 1$

$$(4.160) \quad \int |S_N(t, dw^1)|^{p''} \, G(t, dw^1) \, dP_1(w^1)$$

is uniformly bounded (as $t \downarrow 0$)

A similar result holds for the expansion of $[\det C(t, dw^1, q(t,dw^1,y_0))]^{-1}$, so that finally if $G(t, dw^1) \neq 0$

$$(4.161) \quad \frac{\det C \, (1+t^{1/2} k(t, dw^1) + \ldots + t^N k^{2N}(t, dw^1))}{\det C(t, dw^1, q(t, dw^1, y_0))} =$$

$$1 + \sum_{1}^{2N} d_k(dw^1) \, t^{k/2} + t^{\frac{2N+1}{2}} S'_N(t, dw^1)$$

and moreover for any $p'' > 1$

$$(4.162) \quad \int |S'_N(t, dw^1)|^{p''} \, G(t, dw^1) \, dP_1(w^1)$$

is uniformly bounded.

By (4.71), Theorem 4.17, Proposition 4.20 and Hölder's inequality, it is clear that as $t \downarrow\downarrow 0$, for any $k' \in \mathbb{N}$

$$\int \exp\{-\frac{1}{2}\int_0^1 <\lambda,\partial^2 v^2(0)> ds\}\ d_k(dw^1) 1_{G(t,dw^1)\neq 1}\ dP_1(w^1) = o(t^{k'})$$

so that we may replace $G(t, dw^1)$ by 1 in all the terms where it appears.

Using (4.91), (4.158), (4.161), (4.162) we see that to prove (4.147), we only need to show that for any $l \in \mathbb{N}$

(4.163) $\quad \int \exp\{-\frac{1}{2}\int_0^1 <\lambda,\partial^2 v^2(0)>\}\ d_{2l+1}(dw^1)\ dP_1(w^1) = 0.$

Note that P_1 is invariant under the mapping $w^1 \to -w^1$. Now it is not hard to check that the "degree" of $d_{2l+1}(dw^1)$ in the variable dw^1 is odd, so that $d_{2l+1}(-dw^1) = -d_{2l+1}(dw^1)$.

(4.163) follows. □

j) A second split

As pointed out in Remark 1, the projection operator $\bar{P}_2$ is not easy to calculate, and so the differentials $\partial^i v^2(0)$ are also difficult to obtain.

We will now briefly show how to use another description of the diffusion so that the split becomes trivial.

Assume that a is a standard Brownian motion. We define w by the relation

$$(4.164) \quad a_t = w_t + \int_0^t \alpha\lambda \, ds$$

$$\alpha_t = \int_0^t \Omega(\lambda, a) \, ds$$

Of course we will consider the new probability measure

$$(4.165) \quad dP'_t(a) = \exp\left\{ \int_0^1 <\alpha\lambda - \frac{\lambda}{\sqrt{t}}, \delta\, a> - \frac{1}{2}\int_0^1 |\alpha\lambda - \frac{\lambda}{\sqrt{t}}|^2 \, ds \right\} dP(a)$$

so that under $dP'_t(a)$, $w_t + \int_0^t \frac{\lambda}{\sqrt{t}} ds$ is Brownian martingale, and so $\psi_s^t(\sqrt{t}\, dw + \lambda ds, .)$ has the right law.

Of course $\psi_s^t(\sqrt{t}\, dw + \lambda\, ds, .)$ is in fact a function of da. We now can apply to such a function the technique we used for dw, i.e. split the Brownian measure dP(a).

If $\bar{a}_t$ is such that $\bar{a}_0 = 0$, $\dot{\bar{a}} \in H$, define v_t by the relation

(4.166) $\quad \dot{\bar{a}}_t = v_t + \bar{\alpha}\,\lambda$

$$\bar{\alpha}_s = \int_0^t \Omega(\lambda,\bar{a})\,ds$$

Consider the differential equation

(4.167) $\quad du = Y_i(u)\,[\lambda + \ell vds]$

$\quad\quad u(0) = u_0$

To obtain $\dfrac{du}{d\ell}\,\ell=0$, we use (2.2), i.e. if

$$\theta\!\left(\frac{du}{d\ell}\right) = \bar{\theta}\;,\;\; \omega\!\left(\frac{du}{d\ell}\right) = \bar{\omega}\,,\;\text{we get at}\;\; \ell = 0$$

(4.168) $\quad d\bar{\theta} = (v + \bar{\omega}\,\lambda)\,ds\;;\;\;\bar{\theta}'(0) = 0$

$\quad\quad d\bar{\omega} = \Omega(\lambda,\bar{\theta})\,ds\;;\;\;\bar{\alpha}'(0) = 0$

From (4.166), (4.168), we see that

(4.169) $\quad \bar{\theta}_t = \bar{a}_t$

$\quad\quad \bar{\omega}_t = \bar{\alpha}_t$

We then find that

$$\pi^*\frac{\partial\psi}{\partial v}1(\lambda\,ds,u_0)\;\frac{\partial v}{\partial a}(\bar{a}) = u_1\,\bar{a}_1$$

The new matrix C is exactly the identity in $T_{y_0} M$. The split of H will be the trivial

(4.170) $\quad \overline{H}_1 = \{\dot{a} \in H \; ; \; \int_0^1 \dot{a} \, ds = 0\}$

$\overline{H}_2 = \{cst\}$

$\overline{P}_1$ will be the Brownian bridge measure Q, and $\overline{P}_2$ the Gaussian measure on the constants.

The projection operator on H_2 is given by

(4.171) $\quad \dot{a} \in H \to \int_0^1 \dot{a} \, ds \in \overline{H}_2$

By differentiating the relation

(4.172) $\quad \pi \, \psi_1^t(\sqrt{t} \, dw(a_1) + v^2(a_2) \quad ds, u_0) = y_0$

we will get at $t = 0$

(4.173) $\quad M_n(da_1) + \partial^n a_2 = 0$

where $M_n(da_1)$ is already known by recursion.

Another expansion of $p_t(x_0, y_0)$ can then be obtained.

k) Expansion of $p_t(x_0,x_0)$

We now show how Theorem 4.21 permits us to obtain as least the coefficient of t in the Taylor expansion of $(\sqrt{2\pi t})^n p_t(x_0,x_0)$, which gives the Minakshishundaram-Pleijel expansion of $p_t(x_0,x_0)$ [74]. This coefficient was obtained in Mc Kean-Singer [49].

Of course if $x_0 = y_0$, $\lambda = 0$. H_1, H_2 coincide with $\overline{H}_1$, $\overline{H}_2$ in (4.170). P_1 is exactly Q. The projection on H_2 is given by (4.171).

An obvious computation shows that in (4.146)

(4.174) $\quad d_1 = -E_3 - c_1$

$\qquad d_2 = -c_2 - E_4 + E_3 c_1 + \dfrac{E_3^2}{2} + c_1^2$

A trivial (non probabilistic !) computation shows that

(4.175) $\quad E_3 = 0$

$\qquad E_4 = \dfrac{1}{8} |\partial^2 v^2(0)|^2$

If

(4.176) $\quad \theta_s^{(1)} = \theta(\partial \psi_s^t(\sqrt{t}\, dw^1 + v^2(t)\, ds,\, u_0))$

$\qquad \omega_s^{(1)} = \omega(\partial \psi_s^t(\sqrt{t}\, dw^1 + v^2(t)\, ds,\, u_0))$

using equation (2.2) and (4.86), we easily find that at $t = 0$

(4.177) $\quad \theta_s^{(1)} = w_s^1 \qquad \partial v^2 = 0$

$\qquad \omega_s^{(1)} = 0$

If $(\theta^{(2)}, \omega^{(2)}) = (\partial \theta^{(1)}, \partial \omega^{(1)})$, using (4.87), we find that at $t = 0$

(4.178) $\quad \theta_s^{(2)} = 0 \qquad \partial^2 v^2(0) = -2 \theta(\cdot b')(u_0)$

If $u_s^t = \psi_s^t(\sqrt{t}\, dw^1 + v^2\, ds,\, u_0)$, we now compute

(4.179) $\quad \overline{C}(t, dw^1) = [u_1^t]^{-1}\, C(t, dw^1,\, q(t, dw^1, x_0))\, u_1^t$

which is a linear mapping from R^n into R^n and has the same determinant as $C(t, dw^1, x_0)$.

If $X \in R^n$, $\overline{C}(t, dw^1)\, X$ is equal to $\overline{\theta}_1^{(0)}$, where $\overline{\theta}^{(0)}$ is obtained by solving the differential equation

(4.180) $\quad d\overline{\theta}^{(0)} = (X + t\overline{\nabla . b}\,(\overline{\theta}^{(0)}))\, ds + \overline{\omega}^{(0)}(\sqrt{t}\, dw^1 + v^2\, ds)\,;\; \overline{\theta}^{(0)}(0) = 0$

$\qquad d\overline{\omega}^{(0)} = \Omega(\sqrt{t}\, dw^1 + (v^2 + t\, \theta(b'))\, ds,\, \overline{\theta}^{(0)})\,;\; \overline{\omega}^{(0)}(0) = 0$

so that at $t = 0$

(4.181) $\quad \overline{\theta}_s^{(0)} = sX \qquad \overline{\omega}_s^{(0)} = 0.$

We now differentiate (4.180) with respect to $\sqrt{t}$. We get

(4.182) $\quad d\overline{\theta}^{(1)} = \partial \ [t \ \overline{\nabla b}(\overline{\theta}^{(0)})]ds + \overline{\omega}^{(0)}(dw^1 + (\partial \ v^2)ds)$

$\qquad\qquad\qquad + \overline{\omega}^{(1)}(\sqrt{t} \ dw^1 + v^2 ds); \ \overline{\theta}^{(1)}(0) = 0$

$\quad d\overline{\omega}^{(1)} = - [\omega^{(1)}, \Omega(\sqrt{t} \ dw^1 + (v^2 + t\theta(b')) \ ds, \ \overline{\theta}^{(0)})]$

$\qquad\qquad + \overline{\nabla_{u\theta^{(1)}} R}(\sqrt{t} \ dw^1 + (v^2 + t\theta(b')) \ ds, \ \overline{\theta}^{(0)})$

$\qquad\qquad + \Omega(\omega^{(1)}(\sqrt{t} \ dw^1 + (v^2 + t\theta(b')) \ ds) + dw^1 + (\partial \ v^2 + \partial \ t\theta(b')) ds, \overline{\theta}^{(0)})$

$\qquad\qquad + \Omega(\sqrt{t} \ dw^1 + (v^2 + t\theta(b')) \ ds, \ \omega^{(1)} \overline{\theta}^{(0)} + \overline{\theta}^{(1)})$

$\overline{\omega}^{(1)}(0) = 0$

Here $\overline{\nabla.R}$ is the equivariant representation of $\nabla.R$ (R is the curvature tensor). Of course, we still use the convention given after (2.2). Now

(4.183) $\quad \omega^{(1)} t\theta(b') + \partial \ t \ \theta(b) = 2\sqrt{t} \ \theta(b') + t \ \overline{\nabla b}(\theta^{(1)})$

At $t = 0$, we get

(4.184) $\quad \overline{\theta}^{(1)} = 0 \ ; \ \overline{\omega}_s^{(1)} = \int_0^s \Omega(dw_h^1, hX)$

Differentiating the first line of (4.182) at $t = 0$, we get

(4.185) $\quad d\overline{\theta}^{(2)} = \partial^2(t \ \overline{\nabla b} \ (\overline{\theta}^{(0)}))ds + 2\overline{\omega}^{(1)} \ dw^1 \ ; \ \overline{\theta}^{(2)}(0) = 0$

so that

(4.186) $\quad \overline{\theta}_1^{(2)} = 2 \int_0^1 \overline{\nabla b} \ (sX) \ ds + 2 \int_{0 \leq h \leq s \leq 1} \Omega(dw_h^1, hX) \ dw_s^1$

Now obviously

(4.187)
$$\partial \det \overline{C} = (\det \overline{C})(\operatorname{Tr} \overline{C}^{-1} \partial \overline{C})$$
$$\partial^2 \det \overline{C} = (\partial \det \overline{C})(\operatorname{Tr} \overline{C}^{-1} \partial \overline{C}) + (\det \overline{C}) \partial (\operatorname{Tr} \overline{C}^{-1} \partial \overline{C})$$
$$\partial^3 \det \overline{C} = (\partial^2 \det \overline{C})(\operatorname{Tr} \overline{C}^{-1} \partial \overline{C}) + 2(\partial \det \overline{C}) \partial (\operatorname{Tr} \overline{C}^{-1} \partial \overline{C})$$
$$+ (\det \overline{C}) \partial^2 \operatorname{Tr}(\overline{C}^{-1} \partial \overline{C})$$

so that using (4.181), (4.183), (4.184), (4.186), we get

(4.188) $\quad c_1 = 0$

$$c_2 = \frac{\operatorname{Tr}}{2!}(\partial^2 \overline{C}) = \frac{\operatorname{div} b(x_0)}{2} + \sum_1^n \int_{0 \leq h \leq s \leq 1} <\Omega(dw_h^1, h\, e_i)\, dw_s^1,\, e_i>$$

Now, if J is the equivariant representation of the Ricci tensor S

(4.189)
$$\sum_1^n \int_{0 \leq h \leq s \leq 1} <\Omega(dw_h^1, h\, e_i)\, dw_s^1,\, e_i> = -\int_{0 \leq h \leq s \leq 1} J(h\, dw_h^1,\, dw_s^1)$$
$$= \int_0^1 h\, J(dw_h^1,\, w_h^1)$$

Since J is symmetric, we can integrate by parts and obtain that (4.189) is equal to

(4.190) $\quad -\frac{1}{2} \int_0^1 J(w_s^1, w_s^1)\, ds.$

so that

(4.191) $\quad c_2 = \dfrac{\operatorname{div} b(x_0)}{2} - \dfrac{1}{2} \int_0^1 J(w_s^1, w_s^1)\, ds.$

Let K be the scalar curvature, i.e. the trace of the Ricci tensor. From (4.174), (4.175), (4.178), (4.191), we find

(4.192) $\quad d_2 = -c_2 - E_4 = -\dfrac{\text{div } b(x_0)}{2} - \dfrac{|b|^2(x_0)}{2} + \dfrac{1}{2}\int_0^1 J(w_s^1, w_s^1)\, ds.$

Clearly

(4.193) $\quad \int\left[\int_0^1 \dfrac{1}{2} J(w_s^1, w_s^1)\, ds\right] dP_1(w^1) = \dfrac{1}{2}\int\left[\int_0^1 J_{ii}\, w^{1,i}(s)\, w^{1,i}(s)\, ds\right] dP_1(w^1) =$

$$= \dfrac{K}{2}\int_0^1 s(1-s)\, ds = \dfrac{K}{12}$$

Using (4.147), we find that

(4.194) $\quad p_t(x_0, x_0) = \dfrac{1}{(\sqrt{2\pi t})^n}\, [1 + (\dfrac{K}{12} - \dfrac{\text{div } b(x_0)}{2} - \dfrac{|b|^2(x_0)}{2})\, t + \ldots]$

This is exactly the result in Mc Kean-Singer [49] (with the observation that the scalar curvature in [49] is equal to $\dfrac{1}{2} K$).

Of course tensor-product arguments like in Mc Kean-Singer [49] can give part of the terms in t^2.

Our method gives a systematic way of constructing the coefficients of the expansion of $p_t(x_0, x_0)$. To do this, we need to

- compute (d_i) in terms of (E_i), (c_i)
- evaluate (E_i) in terms of the $(\partial^i v^2)$
- compute $(\partial^i v^2)$ by differentiating as many times as needed the equation of $\theta^{(1)}$ (4.86).
- compute (c_i) by differentiating the equations of $\bar{\theta}^{(0)}$, using the evaluations of $(\theta^{(i)}, \omega^{(i)})$, and the continuation of (4.182)
- evaluate the expectations of (d_i). This is of course non trivial.

V. THE HYPOELLIPTIC CASE : TWO CONJECTURES.

In this section, we formulate two conjectures on the asymptotic behavior as $t \downarrow\downarrow 0$ of the semi-group associated to an hypoelliptic operator of the form :

(5.1) $$\mathcal{L}' = X_0 + \frac{1}{2} \sum_1^m X_i^2$$

We will assume that assumption H2 of Definition 1.9 is verified, so that the Malliavin covariance matrix is invertible on the considered deterministic paths.

In a), we introduce the main assumptions and notations. In b), we do some remarks on the stochastic calculus for processes which are not semi-martingales. It turns out that in general, for the new measure P_1, w^1 is a semi-martingale on $[0,1[$, but not on $[0,1]$. We show that for our purposes, a convenient stochastic calculus can still be developed on $[0,1]$.

In c), we formulate two conjectures which naturally extend Theorems 4.16 and 4.21. Since we need to verify the integrability of some exponential random variables, we discuss various results on quadratic forms acting on certain Gaussian bridges.

In d), the conjectures are proved for the elliptic case, just as in section 4.

In e), we find that if M is the Heisenberg group, the equivalent of $P_t(x_0,y_0)$ obtained via conjecture n°2 is exactly what has been found in Gaveau [8] - [32].

Let us insist on the fact that in the general case, the two conjectures could prove to be wrong.

a) <u>Assumptions and notations</u>.

The assumptions and notations are the same as in Section 1.

We will assume that assumption H2 of Definition 1.9 is verified <u>at every</u> $x_0 \in M$.

$X_0(x)$ is another C^∞ vector field on M which has the same properties as $X_1,\ldots,X_m$.

Ω is the probability space $\mathscr{C}(R^+;R^m)$ and P denotes the Wiener measure on Ω.

Consider the stochastic differential equation on (Ω,P) for $t > 0$

(5.2) $$dx = tX_0(x)ds + \sqrt{t}\, X_i(x).dw^i$$

$$x(0) = x_0$$

(5.2) defines a flow of diffeomorphisms $\psi_s^t(\sqrt{t}\,dw,.)$ so that the solution of (5.2) is $\psi_s^t(\sqrt{t}\,dw,x_0)$. Of course $\psi_s^t(\sqrt{t}\,dw,x_0)$ is jointly continuous in (t,s,x_0).

x_0, x are now fixed as in Section 1 e), and verify H3. Because of H2, if $x \neq x_0$, H3 is necessarily verified. If $x = x_0$, $\mathcal{L}$ must be elliptic at x_0.

We also do the following assumption :

H4 : We assume that $h \in H$ such that (1.42) holds is <u>unique</u>, and moreover that x_0, x are not <u>conjugate</u>, i.e. there is no $v \in H_1 \neq 0$ such that (1.65) holds (also see (1.67)).

Since h is a minimum of I on $K_x^{x_0}$, the quadratic form $I''(h)$ in (1.63) is positive definite on H_1.

P_1, P_2 are the Gaussian cylindrical measures on H_1 and H_2.

Using Theorem 1.19, we know that on (Ω,P), if

(5.3) $$\dot{w}_s^2 = \rho([C_1^{h,x_0}\,{}^{-1}]\int_0^1 \phi_s^{h*-1}\,X_i(x_0)dw^i)$$

$$w_t^1 = w_t - \int_0^t \dot{w}_s^2\,ds$$

then w^1 and $\dot{w}^2$ are independent, the "law" of dw^1 is P_1 and the law of $\dot{w}^2$ is P_2.

b) <u>Remarks on the semi-martingale property</u>.

Let $\{F'_s\}_{s \geq 0}$ be the enlarged filtration associated to

(5.4) $\quad F'_s = \mathcal{B}(w_h | h \leq s) \vee \mathcal{B}(\int_0^1 (\phi_s^{h*-1} X_i)(x_0) dw^i)$

It must be pointed out that w is a semi-martingale with respect to $\{F'_s\}_{s \geq 0}$ on every time interval $[0,\eta]$ with $\eta < 1$, but that except in the case where $\mathcal{L}$ is elliptic everywhere, it is in general <u>not</u> a semi-martingale on $[0,1]$. To see this, let D_s^{h,x_0} be the linear mapping from $T^*_{x_0}$ into $T_{x_0} M$

(5.5) $\quad p \in T^*_{x_0} M \to D_s^{h,x_0} p = \int_s^1 <(\phi_r^{h*-1} X_i)(x_0), p> (\phi_r^{h*-1} X_i)(x_0) dr$

For $s < 1$, D_s^{h,x_0} is invertible. Indeed, if $x = x_0$ $\mathcal{L}$ is elliptic at x_0 and D_s^{h,x_0} is obviously invertible. If $x \neq x_0$, since $\mathcal{H}$ is preserved under ψ, h is a.e.$\neq 0$ on $[0,1]$ (otherwise it would be 0 everywhere), and by H2, for $s < 1$, D_s^{h,x_0} is invertible.

From Chaleyat-Jeulin [75], we easily find that w is a semi-martingale with respect to $\{F'_s\}_{s \geq 0}$ on $[0,1]$ if :

(5.6) $\quad \int_0^1 (\text{Tr } [D_s^{h,x_0}]^{-1})^{1/2} ds < +\infty$.

Now, under H2, $\text{Tr } [D_s^{h,x_0}]^{-1} \sim (1-s)^{-m}$ where $m \geq 2$ if $\mathcal{L}$ is not elliptic at x. If $\mathcal{L}$ is not elliptic at x, (5.6) cannot be verified.

We can give an equivalent formulation of the previous result. Namely under P_1, w^1 is a semi-martingale (with respect to its own filtration) on $[0,\eta]$, with $\eta < 1$, but in general is not a semi-martingale on $[0,1]$.

Since time reversal applies obviously on (5.2), we find that if w'^1 is defined by :

(5.7) $\qquad w'^1_s = w^1_1 - w^1_{1-s} \qquad (0 \le s \le 1)$

w'^1 is a semi-martingale on $[0,\eta]$ (for $\eta < 1$) but in general is not a semi-martingale on $[0,1]$.

However, certain stochastic integrals on $[0,1]$ are still well-defined. To see this we could use time reversal, i.e. note that since w^1 and w'^1 are semi-martingales on $[0,\frac{1}{2}]$, certain stochastic integrals with respect to w^1 on $[0,1]$ can also be expressed as sums of products of well-defined stochastic integrals with respect to w^1 and w'^1 on $[0,\frac{1}{2}]$

We will here use a more direct procedure. Namely for $k \in \mathbb{N}$, let $E_k^{j_1,\ldots,j_k}(t_1,\ldots,t_k)$ be a bounded measurable function on $\mathbb{R}^k$.

Recall that w is a semi-martingale with respect to $\{F'_s\}_{s \ge 0}$ on $[0,1[$. If $s < 1$, we may calculate the stochastic integral

(5.8) $\qquad \displaystyle\int_{0 \le t_1 \le \ldots \le t_k \le s} E_k^{j_1,\ldots,j_k}(t_1,\ldots,t_k)\, dw^{1,j_1}_{t_1} \ldots dw^{1,j_k}_{t_k}$

on the probability space (Ω,P) of the Brownian motion, i.e. express w^1 as a function of w as in (5.3), and use the fact that since w^1 is a semi-martingale with respect to $\{F'_s\}_{s\geq 0}$, we may as well obtain (5.8) as a stochastic integral with respect to w. Of course (5.8) will still be adapted to $\{\mathcal{B}(w^1_h \mid h\leq s)\}_{s\geq 0}$.

Using (5.3), we see that the terms $\int_0^1 \phi_s^{h*-1} X_i(x_0) \cdot dw^i$ will factor in (5.8), so that (5.8) can in fact be expressed in terms of standard stochastic integrals with respect to w, which have P a.s. limits as $s \uparrow\uparrow 1$.

We then find that P_1 a.s., as $s \uparrow\uparrow 1$, the limit of (5.8) exists and will be noted as

(5.9) $$\int_{0\leq t_1\ldots \leq t_k \leq 1} E_k^{j_1,\ldots,j_k}(t_1,\ldots,t_k)\, dw_{t_1}^{1,j_1},\ldots,dw_{t_k}^{1,j_k}$$

(5.9) is of course an extended stochastic integral.

All the stochastic integrals which will later appear are of this type.

c) - <u>Two conjectures</u>.

We will now formulate one conjecture, which so far, we have not get proved, for lack of basic estimates, like the <u>global</u> estimates of Varadhan [69], [70], and Azencott [6], [8] on diffusions on <u>complete</u>

Riemannian manifolds. The fact that the estimates of Varadhan [69], [70], Molchanov [54], Azencott [6], [8] blow up when the Riemannian manifold is not complete is a clear indication that indeed a global control of the diffusions is needed in order to obtain the behavior of $p_t(x_0,y_0)$ along the path of minimal action.

$\ell_s^i \in T_x M$ is defined by

(5.10) $\quad \ell_s^i = \phi_1^{h*}(\phi_s^{h*-1} X_i)(x_0).$

Similarly, we set

(5.11) $\quad p_0' = \phi_1^{h*}(x_0) p_0 \in T_x^* M.$

From Theorem 1.17, we know that

(5.12) $\quad h_s^i = <p_0', \ell_s^i>.$

If $q \in T_x^* M$, $<q, \ell_s>$ is the vector of R^m

(5.13) $\quad (<q, \ell_s^1> \ldots <q, \ell_s^m>).$

We also define $\psi_\cdot^t(\sqrt{t}\, dw, q, \cdot)$ to be the flow associated to the stochastic differential equation

(5.14) $\quad dx = t\, X_0(x) ds + X_i(x)(\sqrt{t}\, dw^i + <q, \ell>^i ds)$
$\quad\quad\quad x(0) = x_0'$

Of course $\psi_s^t(\sqrt{t}\,dw,q,x_0')$ is still jointly continuous in (t,s,q,x_0'). By proceeding as in (4.26), $\psi_s^t(\sqrt{t}\,dw^1,q,x_0')$ is also well-defined for $s \in [0,1]$.

Recall that ρ has been defined in Definition 1.18.

<u>Definition 5.1</u> : C' is the linear mapping from $T_x^* M$ in $T_x M$

(5.15) $\quad p \in T_x^* M \to C'p = \phi_1^{h*}(x_0)\, C_1^{h,x_0}\, \widetilde{\phi_1^{h*}(x_0)}\, p$

$\dfrac{\partial \psi_1^t}{\partial v}(\sqrt{t}\,dw^1,q,x_0)$ is the linear mapping :

(5.16) $\quad v \in H \to \psi_1^{t*}\displaystyle\int_0^1 (\psi_s^{t*-1}(\sqrt{t}\,dw^1,q,\cdot)X_i)(x_0)\, v^i ds \in T_{\psi_1^t(\sqrt{t}\,dw^1,q,x_0)} M$

If $\psi_1^t(\sqrt{t}\,dw^1, p_0'+q, x_0) = x$, $C'(t,dw^1,q)$ is the linear mapping

(5.17) $\quad p \in T_x^* M \to \dfrac{\partial \psi_1^t}{\partial v}(\sqrt{t}\,dw^1, p_0'+q, x_0)\, \rho(\widetilde{\phi_1^{h*}(x_0)}p) \in T_x M$

It is easy to check that :

(5.18) $\quad \dfrac{\partial \psi_1^t}{\partial q}(\sqrt{t}\,dw^1, p_0'+q, x_0) = C'(t,dw^1,q)$

$C'(0,dw^1,0) = C'$

Since C' is invertible, the same discussion as in Section 4.e) applies. Namely if y is close enough to x and if t is small enough, the equation

(5.19) $\quad \psi_1^t(\sqrt{t}\, dw^1,\, p_0' + q,\, x_0) = y$

has one single solution $q(t,dw^1,y)$ in a small enough neighborhood of 0 in $T_x^* M$. Of course, the precise definition of $q(t,dw^1,y)$ is a word for word adaptation of what is done in Section 4.e).

$q(t,dw^1,y)$ is then a smooth function of $(\sqrt{t},y)$. Similarly, we define:

(5.20) $\quad v^2(t,dw^1,y) = \rho(\widetilde{\phi_1^{h*}(x_0)} q(t,dw^1,y))$

In what follows we assume that M is endowed with a Riemannian metric, and so has a volume form dy. $\det C'(t,dw^1,q(t,dw^1,x))$ is then well-defined.

$p_t(x_0,y)dy$ is the law of $\psi_1^t(\sqrt{t}\, dw, x_0)$ under P. Of course $p_t(x_0,\cdot) \in C^\infty(M)$.

In the next statement, we entirely forget about mollifiers. Of course, to make what follows precise, we should introduce them as in Theorem 4.16.

We will also use the notation (4.56).

We then have a first conjecture:

Conjecture 5.2 : As $t \downarrow\downarrow 0$.

$$(5.21) \quad p_t(x_0,x) \equiv \frac{[\det C']^{1/2}}{(\sqrt{2\pi t})^n} \int_\Omega \frac{\exp -\int_0^1 \frac{|h+v^2(t,dw^1,x)|^2 ds}{2t}}{\det C'(t,dw^1,q(t,dw^1,x))} dP_1(w^1)$$

Of course, it must be pointed out that the set where $q(t,dw^1,x)$ is not well-defined (or where $\det C'(t,dw^1,q(t,dw^1,x))$ could be 0) has P_1-measure $o(t^k)$ (for any $k \in N$).

The "proof" of (5.21) is exactly the same as the proof of Theorem 4.16, once results similar to the results of Section 3 and 4 have been established. This is an important proviso, which could eventually make the conjecture fail.

$E_k(dw^1)$, $c_k(dw^1)$, $d_k(dw^1)$ are defined in exactly the same way as in (4.143), (4.145), (4.146) (with C' replacing C). The only new fact is that if $\mathcal{L}$ is not elliptic, there are in general <u>extended</u> stochastic integrals on [0,1] in the sense of Section 5. b). Of course, they still are in all the $L_p(\Omega,P_1)$ ($1 \le p < +\infty$).

Before stating the next conjecture, we still need an intermediary result.

<u>Theorem 5.3</u> : There is $\mu > 1$ such that :

$$(5.22) \quad \int_\Omega \exp\{\frac{\mu}{2} < p_0, \int_{0 \le s \le t \le 1} [(\phi_s^{h*-1} X_i)(x_0), (\phi_t^{h*-1} X_j)(x_0)] dw^{1,i}(s) dw^{1,j}(t) >\}$$
$$dP_1(w^1) < +\infty$$

Proof: Set

(5.23) $\bar{K}^{i,j}(s,t) = -\frac{1}{2} <p_0,[\phi_s^{h*-1}X_i(x_0),(\phi_t^{h*-1}X_j)(x_0)]>$ if $s \leq t$

$\frac{1}{2} <p_0,[\phi_s^{h*-1}X_i(x_0),\phi_t^{h*-1}X_j(x_0)]>$ if $s > t$

$\bar{K}^{i,j}(s,t)$ defines a symmetric kernel on $H \times H$, and so a symmetric Hilbert-Schmidt operator $\bar{K}$ on H. We may express $\bar{K}$ in the form

(5.24) $\bar{K} = \sum_{1}^{+\infty} \lambda_k <f_k,.> f_k$

where $\{\lambda_k\}_{k\in\mathbb{N}}$ are the real eigenvalues of $\bar{K}$, and $\{f_k\}_{k\in\mathbb{N}}$ the corresponding unit eigenvectors.

The series in the r.h.s. of (5.24) converges to $\bar{K}$ in $H \otimes H$.

Note that in (5.22), the Stratonovitch stochastic integrals are in fact Itô integrals (because $[(\phi_s^{h*-1}X_i)(x_0),(\phi_s^{h*-1}X_i)(x_0)] = 0$).

We now proceed as Hida [71] does for the standard Brownian motion.

Using (5.1) and the results of Section 5.b), we may express

$$\int_{0\leq s\leq t\leq 1} \bar{K}^{i,j}(s,t) \delta w_s^{1,i} \delta w_t^{1,j}$$

in terms of w. We obtain, using (5.3) and writing C instead of C_1^{h,x_o}

(5.25) $\int_{0\leq s\leq t\leq 1} \bar{K}^{i,j}_{(s,t)} \delta w^i_s \delta w^j_t - \int_{0\leq s\leq t\leq 1} \bar{K}^{i,j}_{(s,t)} <(\phi_s^{h*-1} X_i),$

$$C^{-1} \int_0^1 (\phi_u^{h*-1} X_k) \delta w_u^k > \delta w_t^j$$

$-\int_{0\leq s\leq t\leq 1} \bar{K}^{i,j}_{(s,t)} \delta w^i_s <\phi_t^{h*-1} X_j, C^{-1} \int_0^1 (\phi_u^{h*-1} X_k) \delta w_u^k>$

$+\int_{0\leq s\leq t\leq 1} \bar{K}^{i,j}_{(s,t)} <\phi_s^{h*-1} X_i, C^{-1} \int_0^1 \phi_u^{h*-1} X_k \delta w_u^k><\phi_t^{h*-1} X_j,$

$$C^{-1} \int_0^1 \phi_v^{h*-1} X_\ell \delta w_v^\ell>$$

Of course, in (5.25) all the terms are either standard Itô integrals or products of standard Itô integrals.

We claim that (5.25) (which is in $L_2(\Omega,P)$) depends continuously on $K \in H \otimes H = L_2([0,1]^2 ; R^m \otimes R^m)$. This statement is clear for the first and the last term in (5.25). We now consider the second term. From Cauchy-Schwarz's and Burkholder - Davis - Gundy's inequalities, if (e_ℓ) is a basis of $T_{x_o} M$, and (e^ℓ) the dual basis in $T^*_{x_o} M$

- 195 -

$$(5.26) \quad E \left| \left[\int_{0 \leq s \leq t \leq 1} \bar{K}^{i,j}(s,t) <\phi_s^{h*-1} X_i, e^\ell> \delta w_t^j \right] < C^{-1} \int_0^1 \phi_u^{h*-1} X_k \delta w^k, e_\ell> \right|^2$$

$$\leq C \left[E \left| \int_{0 \leq s \leq t \leq 1} \bar{K}^{i,j}(s,t) <\phi_s^{h-1} X_i, e^\ell> \delta w_t^j \right|^4 \right]^{1/2}$$

$$\leq C' \int_{0 \leq t \leq 1} dt \left| \int_{0 \leq s \leq t} |\bar{K}^{i,j}(s,t)| ds \right|^2 \leq C'' \int_{0 \leq s \leq t \leq 1} |\bar{K}^{i,j}(s,t)|^2 ds dt$$

From (5.26), we see that the second term in (5.25) depends continuously on $\bar{K} \in H \otimes H$. Using time reversal, this is also the case for the third term in (5.25).

Since the series in (5.24) converges to $\bar{K}$ in $H \otimes H$, we can write that

$$(5.27) \quad <p_0, \int_{0 \leq s \leq t \leq 1} [(\phi_s^{h*-1} X_i)(x_0), (\phi_t^{h*-1} X_j)(x_0)] \delta w_s^{1,i} \delta w_t^{1,j}>$$

$$= -2 \sum_1^{+\infty} \lambda_k \int_{0 \leq s \leq t \leq 1} <f_k(s), \delta w_s^1> <f_k(t), \delta w_t^1>$$

$$= - \sum_1^{+\infty} \lambda_k \left[\left(\int_0^1 <f_k(s), \delta w_s^1> \right)^2 - \int_0^1 |f_k(s)|^2 ds \right]$$

where the series in (5.27) converges in $L_2(\Omega, P_1)$.

Now since $\bar{P}_2$ has finite dimensional range, $\bar{P}_2$ is Hilbert-Schmidt so that

$$(5.28) \quad \sum_1^{+\infty} |\bar{P}_2 f_k|^2 < +\infty$$

Since $\{\lambda_k\}$ is a bounded sequence, we have :

(5.29) $\sum_{1}^{+\infty} |\lambda_k| \, |\bar{P}_2 f_k|^2 < +\infty$

(5.27) is then equal to :

(5.30) $-\sum_{1}^{+\infty} \lambda_k [(\int_0^1 <f_k(s), \delta w_s^1>)^2 - \int_0^1 |\bar{P}_1 f_k|^2 \, ds] + \sum_{1}^{+\infty} \lambda_k \int_0^1 |\bar{P}_2 f_k|^2 \, ds$

$= -\sum_{1}^{+\infty} \lambda_k [(\int_0^1 <\bar{P}_1 f_k, \delta w>)^2 - \int_0^1 |\bar{P}_1 f_k|^2 ds] + \sum_{1}^{+\infty} \lambda_k \int_0^1 |\bar{P}_2 f_k|^2 ds.$

Let λ_k', g_k be the eigenvalues and the unit eigenvectors of the restriction $\bar{K}_{H_1}$ of the quadratic form $\bar{K}$ to H_1. We claim that

(5.31) $\sum_{1}^{+\infty} \lambda_k [(\int_0^1 <\bar{P}_1 f_k, \delta w^1>)^2 - \int_0^1 |\bar{P}_1 f_k|^2 ds] = \sum_{1}^{+\infty} \lambda_k' ((\int_0^1 <g_k, \delta w^1>)^2 - 1)$

Recall that the l.h.s. is in fact a converging series in $L_2(\Omega, P_1)$, which depends continuously on $\bar{K} \in H \otimes H$. A martingale argument shows that the r.h.s. of (5.31) also defines a converging series in $L_2(\Omega, P_1)$ which depends continuously on $\bar{K}_{H_1}$ for the corresponding Hilbert-Schmidt norm.

If $\bar{K}$ has finite dimensional range, both sides of (5.31) are trivially equal to

(5.32) $<\bar{K} dw^1, dw^1> - \text{Tr } \bar{K}_{H_1}$

Since finite dimensional range operators are dense in $H \otimes H$ (see Reed-Simon [72], Chap. VI) (5.31) is still true for a general $\bar{K}$.

(5.30) is then equal to

(5.33) $\quad -\sum_{1}^{+\infty} \lambda'_k (|\int_0^1 <g_k, \delta w^1>|^2 - 1) + \sum_{1}^{+\infty} \lambda_k \int_0^1 |\bar{P}_2 f_k|^2 ds$

By H4, we know that $I + \bar{K}_{H_1}$ is positive definite. Since $\bar{K}_{H_1}$ is compact, we even know that

(5.34) $\quad I + \bar{K}_{H_1} \geq \alpha > 0.$

and so for $\mu > 1$ so that $\mu-1$ is small enough

(5.35) $\quad I + \mu \bar{K}_{H_1} > 0$

Now by the result in Simon [59] p. 30, since $\mu \bar{K}_{H_1}$ is Hilbert-Schmidt:

(5.36) $\quad \int_\Omega \exp\{-\frac{\mu}{2} (\sum_1^{+\infty} \lambda'_k (<g_k, dw^1>^2 - 1)\} dP_1(w^1) = \dfrac{1}{[\det_2(I+\mu\bar{K}_{H_1})]^{1/2}}$

where

(5.37) $\quad \det_2(I + \mu\bar{K}_{H_1}) = \prod_1^{+\infty} (1 + \mu\lambda'_k) e^{-\mu\lambda'_k}$

Let $\bar{K}_{H_2}$ be the restriction of $\bar{K}$ to H_2.

From (5.27)-(5.36), we find that

(5.38) $\int_\Omega \exp\{\frac{\mu}{2} <p_0, \int_{0\leq s\leq t\leq 1} [\phi_s^{h*-1}X_i(x_0),(\phi_t^{h*-1}X_j)(x_0)]dw^{1,i}(s)dw^{1,j}(t)>\} dP_1(w^1) =$

$$\frac{\exp \frac{\mu}{2} \mathrm{Tr}[\bar{K}_{H_2}]}{[\det_2(I+\mu\bar{K}_{H_1})]^{1/2}}$$

The theorem follows. □

<u>Remark 1</u> : It must be pointed out that in general, $\bar{K}$ is not trace-class, which makes the previous computations somewhat complicate .

Recall that $\bar{E}$ has been defined in Definition 1.13. We are now ready to state the second conjecture.

<u>Conjecture 5.4</u> : For any N

(5.39) $p_t(x_0,x) = \frac{1}{\sqrt{2\pi t})^n(\det C')^{1/2}} \exp\{\frac{-\bar{E}(x)}{2t} + <p_0, \int_0^1 (\phi_s^{h*-1}X_0)(x_0)ds>\}$

$[\int_\Omega \exp\{\frac{1}{2} <p_0, \int_{0\leq s\leq s'\leq 1} [(\phi_s^{h*-1}X_i)(x_0),(\phi_{s'}^{h*-1}X_j)(x_0)]dw^{1,i}(s)dw^{1,j}(s')> \}$

$(1 + \sum_1^N d_{2k}(dw^1)t^k) \; dP_1(w^1) + o(t^N)]$

Of course, due to Theorem 5.3, all the integrals exist in the r.h.s of (5.39).

Remark 2 : Using the notations of the proof of Theorem 5.3, we find that if the conjecture 5.4 is verified then as $t \downarrow\downarrow 0$

$$(5.40) \qquad p_t(x_0,x) \sim \frac{1}{(\sqrt{2\pi t})^n (\det C')^{1/2}} \exp \{ \frac{-\bar{E}(x)}{2t} + \int_0^1 <p_0, \phi_s^{h*-1} X_0(x_0)> \, ds \}$$

$$\frac{\exp \frac{1}{2} \operatorname{Tr} \bar{K}_{H_2}}{[\det_2(I+\bar{K}_{H_1})]^{1/2}}$$

d) - The two conjectures : the elliptic case.

Assume here that $\mathcal{L}$ is everywhere elliptic. The Hamiltonian $\mathcal{H}(x,p)$ defines a scalar product in T^*M, and by duality a Riemannian structure on M. Under the assumptions of Section 1, M is then trivially complete for the Riemannian metric.

<u>Theorem 5.5.</u> : If $\mathcal{L}$ is everywhere elliptic, Conjectures 5.2 and 5.4 hold.

<u>Proof</u> : The proof is the same as the proof of Theorems 4.16 and 4.21.
□

Remark 3 : The fact that, say, the term of order 0 found in Conjecture 5.4 is the same as the one which was found in Theorem 4.21 is not absolutely trivial when using a direct method.

e) **A case study : the Heisenberg group.**

In [32],[33], Gaveau has studied the behavior of the semi-group associated to the operator (1.21) as $t \downarrow\downarrow 0$, finding an equivalent for $p_t(x_0,y_0)$ as $t \downarrow\downarrow 0$.

We will now check that the term of order 0 given in Conjecture 5.4 is exactly what has been found in Gaveau [32].

We will follow the notations of Gaveau [32] and Azencott [8].

Here $M = R^3$. An element of M is written $g = (x,y,z)$. X_1, X_2, T are the vectors fields

(5.41) $\quad X_1 = \frac{\partial}{\partial x} + 2y \frac{\partial}{\partial z} \ ; \ X_2 = \frac{\partial}{\partial y} - 2x \frac{\partial}{\partial z}$

$\quad T = \frac{\partial}{\partial z}$

Obviously

(5.42) $\quad [X_1,X_2] = -4T$

Of course (x,y,z) are the coordinates of $g = (x,y,z)$ for the exponential mapping of the Heisenberg group M.

From [8], [32], [33] we know that if $g = (x,y,z)$ is such that $x^2+y^2 \neq 0$, there is one single bicharacteristic curve which connects $e = (0,0,0)$ and g.

$h = (h^1, h^2)$ is the unique element of K_g^e which minimizes I on K_g^e. We then consider the curve $g_t = (x_t, y_t, z_t)$ where $g_t = \phi_t^h(e)$, so that

(5.43) $\quad dg = (X_1(g)h^1 + X_2(g)h^2)ds$

$\qquad g(0) = e.$

On M, we put the Lebesgue measure $dxdydz$ which is the Haar measure of M. Observe that X_1, X_2 have 0 divergence, so that

(5.44) $\quad \det C' = \det C_1^{h,x_0}$

From (1.18), we find:

(5.45) $\quad (\phi_t^{h*-1} X_1)(e) = X_1 - \int_0^t \phi_s^{h*-1}[X_1, X_2](e) h^2 \, ds$

$\qquad \phi_t^{h*-1} [X_1, X_2](e) = [X_1, X_2](e)$

Using (5.45), we get:

(5.46) $\quad (\phi_t^{h*-1} X_1)(e) = X_1 + 4 \int_0^t h^2 ds \, T$

$\qquad (\phi_t^{h*-1} X_2)(e) = X_2 - 4 \int_0^t h^1 ds \, T \; .$

and so from (5.41), (5.43), we get

(5.47) $(\phi_t^{h*-1} X_1)(e) = X_1 + 4 y_t T$

$(\phi_t^{h*-1} X_2)(e) = X_2 - 4 x_t T$.

From (5.47), we easily deduce :

(5.48) $\det C_1^{h, x_0} = 16 \left[\int_0^1 |x_s|^2 ds - \left(\int_0^1 x_s ds \right)^2 + \int_0^1 |y_s|^2 ds - \left(\int_0^1 y_s ds \right)^2 \right]$.

Now H_1 is the set of $v \in H$ such that

(5.49) $\int_0^1 (\phi_s^{h*-1} X_i)(e) v^i ds = 0$

Using (5.47), (5.49) writes

(5.50) $\int_0^1 v^1 ds = 0 \quad ; \quad \int_0^1 v^2 ds = 0$

$\int_0^1 (y_t v^1 - x_t v^2) ds = 0.$

From Gaveau, Azencott [8], [32], [33] we know that

(5.51) $x_t = \alpha(1 - \cos 2\sigma t) + \beta \sin 2\sigma t$

$y_t = \alpha \sin 2\sigma t - \beta(1 - \cos 2\sigma t)$

and that the third component p_0^3 of p_0 is

(5.52) $p_0^3 = \frac{\sigma}{2}$

Here $\sigma \in]-\pi,+\pi[$, and α, β are such that

(5.53) $$\alpha^2 + \beta^2 = \frac{x^2+y^2}{4\sin^2\sigma}$$

A third relation exists between (x,y,z) and σ, but we will not need it.

Clearly

(5.54) $$\frac{1}{2} <P_0, \int_{0\leq s\leq t\leq 1} [\phi_s^{h*-1}X_i, \phi_t^{h*-1}X_j] v^i(s)v^j(t) \, ds \, dt> =$$

$$- \sigma \int_{0\leq s\leq t\leq 1} (v^1(s)v^2(t) - v^2(s)v^1(t)) ds \, dt.$$

Let $\bar{H}_1$ be the Hilbert space :

(5.55) $$\bar{H}_1 = \{v \in H \; ; \int_0^1 v \, ds = 0\}$$

Let Q be the Gaussian cylindrical measure on $\bar{H}_1$. Q is exactly the probability law of the two dimensional Brownian bridge $a = (a^1, a^2)$, with $a(o) = a(1) = 0$.

Set

(5.56) $$\bar{x}_t = x_t - \int_0^1 x_s \, ds, \quad \bar{y}_t = y_t - \int_0^1 y_s \, ds \quad \bar{c} = (\bar{y}, -\bar{x})$$

H_1 is given by :

$$H_1 = \{v \in \bar{H}_1 \; ; <\bar{c}, v> = 0\}.$$

The Gaussian measure P_1 on H_1 is exactly the conditional measure $dQ(|<\bar{c},v> = 0)$.

We must now calculate

(5.57) $\qquad E^{P_1} \exp\{-\sigma \int_0^1 (a^1 \delta a^2 - a^2 \delta a^1)\}$

Consider the symmetric bilinear form on $\bar{H}_1$

(5.58) $\qquad B((v^1,v^2),(v'^1,v'^2)) = 2\sigma \int_{0 \le s \le t \le 1} (v^1(s)v'^2(t) - v^2(s)v'^1(t))ds\,dt$

$\qquad\qquad\qquad\qquad\qquad = <A(v^1,v^2),(v'^1,v'^2)>_{\bar{H}_1}$

The eigenvalues $\{\lambda_k\}$ of A are easily seen to be given by

$\qquad\qquad \lambda_k = \frac{\sigma}{k\pi} \qquad k \in Z,\ k \ne 0$

Since $|\sigma| < \pi$, $|\lambda_k| < 1$ for every $k \in Z$. For $k \in N$, let f_k, f'_k be the unit eigenvectors corresponding to the eigenvalues λ_k and $-\lambda_k$. We may express the Hilbert-Schmidt operator A as

(5.59) $\qquad A = \sum_1^{+\infty} \lambda_k (<f_k,.> f_k - <f'_k,.> f'_k)$

Set

(5.60) $$A^n = \sum_1^n \lambda_k (<f_k,.> f_k - <f'_k,.> f'_k)$$

A^n is symmetric, has a finite dimensional range and its trace is equal to 0. Using finite dimensional calculus, we find that since $dP_1 = dQ(.|<\bar{c},v> = 0)$

(5.61) $$E^{P_1} \exp\{-\tfrac{1}{2} <A_n(a^1,a^2),(a^1,a^2)>\} =$$

$$= \frac{\|\bar{c}\|}{<(I+A_n)^{-1}\bar{c},\bar{c}>^{1/2} [\det(I+A_n)]^{1/2}} =$$

$$= \frac{\|\bar{c}\|}{<(I+A_n)^{-1}\bar{c},\bar{c}>^{1/2} [\det_2(I+A_n)]^{1/2}}$$

Now when $n \to +\infty$, the r.h.s converges trivially to the corresponding expression with A instead of A_n. To prove a similar result for the l.h.s, note that as we saw in (5.27) $<A^n(a^1,a^2),(a^1,a^2)>$ converges P_1 a.s. to $<A(a^1,a^2),(a^1,a^2)>$.
Moreover, replacing A_n by μA_n in (5.61), where μ is >1 and such that $\frac{\mu|\sigma|}{\pi} < 1$, we find that the random variables

$$\exp -\tfrac{1}{2} <A_n(a^1,a^2),(a^2,a^2)>$$

are uniformly integrable for P_1. It is then clear that

(5.62) $$E^{P_1} \exp - \sigma \int_0^1 (a^1 \delta a^2 - a^2 \delta a^1) = \frac{\|\bar{c}\|}{<(I+A)^{-1}\bar{c},\bar{c}>^{1/2} [\det_2(I+A)]^{1/2}}$$

Now using (5.48), we see that

(5.63) $$\|\bar{c}\| = \frac{[\det C]^{1/2}}{4}$$

Moreover

(5.64) $$\frac{1}{[\det_2(I+A)]^{1/2}} = E^Q \exp - \sigma \int_0^1 (a^1 \delta a^2 - a^2 \delta a^1)$$

Since $|\sigma| < \pi$, by a classical result of P. Lévy used by Gaveau [32] (also see Yor [78]) (5.64) is equal to :

(5.65) $$\frac{\sigma}{\sin\sigma}.$$

We now must compute $(I+A)^{-1}\bar{c} = (r^1, r^2) \in \bar{H}_1$.

Set :

(5.66) $$R_t^1 = \int_0^t r_s^1 ds, \quad R_t^2 = \int_0^t r_s^2 ds$$

$$z_t = R_t^1 + i R_t^2.$$

We must have

(5.67) $$\ddot{R}^1 - 2\sigma \dot{R}^2 = \dot{y}$$

$$\ddot{R}^2 + 2\sigma \dot{R}^1 = -\dot{x}$$

so that

(5.68) $\quad \ddot{z} + 2i\sigma\dot{z} = \dot{y} - i\dot{x} = 2\sigma(\alpha - i\beta)e^{-2i\sigma s}$

Since $z(0) = z(1) = 0$, we find

(5.69) $\quad z_t = (i\alpha + \beta)se^{-2i\sigma s} - \dfrac{(i\alpha + \beta)e^{-2i\sigma(s+1)}}{e^{-2i\sigma}-1} + \dfrac{(i\alpha + \beta)e^{-2i\sigma}}{(e^{-2i\sigma}-1)}$

Now

(5.70) $\quad \langle (I+A)^{-1} \bar{c}, \bar{c} \rangle = \int_0^1 (r^1\bar{y} - r^2\bar{x})ds = \int_0^1 (r^1 y - r^2 x)ds$

$\qquad\qquad = \operatorname{Re} \int_0^1 \dot{z}(y + ix)ds$

$\qquad\qquad = \operatorname{Re} \int_0^1 \dot{z}(\beta - i\alpha)e^{2i\sigma s} ds$

An obvious computation shows that

(5.71) $\quad \langle (I+A)^{-1}\bar{c}, \bar{c} \rangle = (\alpha^2 + \beta^2) \dfrac{(\sin\sigma - \sigma\cos\sigma)}{\sin\sigma}$

Using (5.62), (5.63), (5.65), (5.71), Conjecture 5.4 tells us that as $t \downarrow\downarrow 0$

(5.72) $\quad p_t(e,g) \sim \dfrac{1}{(\sqrt{2\pi t})^3} \exp\{-\dfrac{\bar{E}(g)}{2t}\} \dfrac{\sigma}{(4\sin\sigma)(\alpha^2+\beta^2)^{1/2}} \dfrac{(\sin\sigma)^{1/2}}{(\sin\sigma-\sigma\cos\sigma)^{1/2}}$

so that using (5.53)

(5.73) $\quad p_t(e,g) \sim \dfrac{1}{(\sqrt{2\pi t})^3} \exp\{-\dfrac{\bar{E}(g)}{2t}\}\dfrac{1}{2}\dfrac{\sigma}{(\sqrt{x^2+y^2})} [\dfrac{\sin\sigma}{\sin\sigma-\sigma\cos\sigma}]^{1/2}$

This is exactly the result of Gaveau [8]- [32].

REFERENCES

1. Abraham R., Marsden J. : Foundations of Mechanics, 2nd Edition. London : Benjamin / Cummings 1978.

2. Albeverio S. and Høegh-Krohn R. : Topics in Infinite dimensional analysis. In Mathematical Problems in theoretical physics, G. Dell'Antonio ed. Lecture Notes in Physics n° 80, pp. 279-302, Berlin : Springer 1978.

3. Albeverio S., Blanchard Ph., Høegh-Krohn R. : Feynman path integrals and the Trace formula for the Schrödinger operators. Commun Math. Phys. 83, 49-76 (1982).

4. Arnold V. : Méthodes mathématiques de la mécanique classique. Moscou : Mir 1976.

5. Atiyah M., Bott R., Patodi V.K. : On the heat equation and the Index Theorem. Invent. Math., 19, 279-330 (1973).

6. Azencott R. : Grandes déviations et applications. Cours de Probabilité de Saint Flour. Lecture Notes in Math, n° 774, Berlin : Springer 1978.

7. Azencott R. : Développements asymptotiques des semi-groupes de diffusions. To appear.

8. Azencott R., Bellaiche A., Bellaiche C., Bougerol P., Maurel M., Baldi P., Elie L., Granara J. : Géodésiques et diffusions en temps petit. Astérisque 84-85. Société Mathématique de France (1981).

9. Bishop R.L., Crittenden R.J. : Geometry of manifolds. New-York : Acad. Press 1964.

10. Bismut J.M. : Mécanique aléatoire. Lecture Notes in Math n° 866. Berlin : Springer 1981.

11. Bismut J.M. : A generalized formula of Ito and some other properties of stochastic flows. Z. Wahrsch 55, 331-350 (1981).

12   Bismut J.M. : Martingales, the Malliavin calculus and hypoellipticity under general Hörmander's conditions. Z. Wahrsch. 56, 469-505 (1981).

13   Bismut J.M. : The calculus of boundary processes, Ann. Ec. Norm. Sup. To appear (1984)

14   Bismut J.M. : An introduction to the stochastic calculus of variations. In "Stochastic differential systems". M. Kohlmann and N. Christopeit Ed. Lecture notes in Control and Information Sciences n° 43, pp. 33-72, Berlin : Springer 1982.

15   Bismut J.M. : Jump process and boundary processes, Proceedings of the Conference of Katata (1982). To appear (1984).

16   Bismut J.M. : Mécanique aléatoire. Ecole de Probabilités de Saint Flour. p. 1-100. Lecture Notes in Math. n° 929, Berlin : Springer 1982.

17   Davies I., Truman A. : On the Laplace asymptotic expansion of conditional Wiener integrals and the Bender-Wu formula for x^{2N} - anharmonic oscillators. J. Math Phys. 24, 255-266 (1983).

18   Dellacherie C., Meyer P.A. : Probabilités et potentiels, Chap. I. IV, Paris : Hermann 1975, Chap V. VIII , Paris : Hermann 1980.

19   De Witt - Morette C. : Feynman's path integrals. I
Linear and affine transformations II. The Feynman Green's function. Commun Math Phys. 37, 63-81 (1974).

20   De Witt - Morette C. : The semiclassical expansion. Ann. Physics 97, 367-399 (1976).

21   De Witt - Morette C., Maheshwari A., Nelson B. : Path Integration in non relativistic quantum mechanics. Physics Reports 50, 255-372 (1979).

22   Dieudonné J. : Fondements de l'Analyse moderne. Paris : Gauthier-Villars 1965.

23 Donsker M., Varadhan S.R.S. : Asymptotic evaluation of certain Markov expectations for large time. Commun Pure and Appl. Math I, 28, 1-47 (1975), II 29, 279-301 (1976), III 29, 389-461 (1976).

24 Doss H.: Quelques formules asymptotiques pour les petites perturbations de systèmes dynamiques. Ann. Inst. H. Poincaré XVI, 17-28, (1980).

25 Duistermaat J.J. : The light in the neighborhood of a caustic. Séminaire Bourbaki 1976-1977, Exposé N° 490.

26 Duistermaat J.J. : Oscillatory integrals, Lagrange immersions and unfolding of singularities, Commun. Pure and Appl. Math. 27, 207-281 (1974).

27 Elworthy K.D. : Stochastic dynamical systems and their flows. In "Stochastic Analysis", A. Friedman and M. Pinsky ed., 79-95. London:Acad. Press. 1978.

28 Elworthy K.D. : Stochastic methods and differential geometry. In Séminaire Bourbaki. Lecture Notes in Math. n° 901, pp. 95-110. Berlin : Springer 1981.

29 Elworthy K.D., Truman A. : Classical mechanics, the diffusion heat equation and the Schrödinger equation on a Riemannian manifold. J. Math. Phys. 22, 2144-2166 (1981).

30 Elworthy K.D., Truman A. : The diffusion equation and classical mechanics : an elementary formula. In Stochastic Processes in Quantum physics pp. 136-146. S. Albeverio and al ed. Lectures Notes in Physics n° 173, Berlin : Springer 1982.

31 Elworthy K.D. : Stochastic differential equations on manifolds. London Math. Soc. Lecture Notes series n° 70, Cambridge : Cambridge Univ. Press. 1982.

32 Gaveau B. : Principe de moindre action, propagation de la chaleur, estimées sous-elliptiques sur certains groupes nilpotents. Acta Math. 139, 96-153 (1977).

33 Gaveau B : Systèmes dynamiques associés à certains opérateurs hypoelliptiques. Bull.Sc. Math. 102, 203-229 (1978).

34 Haussmann U. : On the integral representation of Ito processes, Stochastics 3, 17-27 (1979).

35 Hörmander L. : Hypoelliptic second order differential equations, Acta Math. 119, 147-171 (1967).

36 Ikeda N., Watanabe S. : Stochastic differential equations and diffusion processes. Amsterdam : North Holland 1981.

37 Itô K., McKean H.P. : Diffusion processes and their sample paths. Grundlehren Math. Wissenschaften Band 125, Berlin : Springer 1974.

38 Jeulin T : Semi-martingales et grossissement d'une filtration. Lecture Notes in Math n° 833, Berlin : Springer 1981.

39 Jeulin T., Yor M. : Sur les distributions de certaines fonctionnelles du mouvement brownien. Séminaire de Probabilités n° 15, pp. 210-226, Lecture Notes in Math. n° 850, Berlin : Springer 1981.

40 Kannai Y. : Off diagonal short time asymptotics for fundamental solutions of diffusion equations. Commun. In partial Diff. Eq. 2, 781-830 (1977).

41 Klingenberg W. : Riemannian Geometry, Berlin : de Gruyter 1982.

42 Kobayashi S., Nomizu K. : Foundations of differential geometry Vol. I : New York : Interscience 1963. Vol. II : Interscience 1969.

43 Kunita H. : On the decomposition of solutions of stochastic differential equations, In "Stochastic Integrals", D. Williams ed., pp. 213-255. Lecture Notes in Math n° 851, Berlin : Springer 1981.

44 Kunita H. : Some extensions of Ito's formulas. Séminaire de Probabilités n° 15, pp. 118-141. Lectures Notes in Math. N° 850, Berlin : Springer 1981.

45 Kusuoka S., Stroock D. : Applications of the Malliavin calculus. Part I, To appear.

46 Malliavin P. : Stochastic calculus of variations and hypoelliptic operators. In "Proceedings of the Conference on Stochastic differential operations of Kyoto (1976)". K. Itô ed. pp. 155-263. Tokyo : Kinokuniya and New York : Wiley 1978.

47 Malliavin P. : Géométrie différentielle stochastique. Montréal : Pressesde l'Université de Montréal 1978.

48 Malliavin P. : Champs de Jacobi stochastiques. C.R.A.S. Paris, Série A, 285, 789-792 (1977).

49 McKean Jr. H. and Singer I.M. : Curvature and the eigenvalues of the Laplacian. J. of differential geometry 1, 43-69 (1967).

50 Menikoff A., Sjöstrand J. : On the eigenvalues of a class of hypoelliptic operators. Math. Annal. 235, 55-85 (1978).

51 Menikoff A., Sjöstrand J. : The eigenvalues of hypoelliptic operators III. The non semi-bounded case. J. Anal. Math. 35, 123-150, (1979).

52 Meyer P.A. : Un cours sur les intégrales stochastiques. Séminaire de Probabilités n° X, pp. 245-400. Lecture Notes in Math N° 511, Berlin : Springer, 1973.

53   Meyer P.A. : Flot d'une équation différentielle stochastique.
     Séminaire de Probabilités n° XV, pp. 103-117. Lecture Notes in
     Mathematics n° 850. Berlin : Springer 1981.

54   Molchanov S.A. : Diffusion processes and Riemannian geometry.
     Russian Math. Surveys 30, 1-63 (1975).

55   Pincus M. : Gaussian processes and Hammerstein integral equations
     Trans. Am. Math. Soc. 134, 193-216 (1968).

56   Rotschild L.P., Stein E.M. : Hypoelliptic differential operators
     and nilpotent groups. Acta Math. 137, 247-320 (1976).

57   Schilder M. : Some asymptotic formulas for Wiener integrals. Trans.
     Am. Math. Soc., 125, 63-85 (1966).

58   Shigekawa I. : Derivatives of Wiener functionals and absolute
     continuity of induced measures. J. Math. Kyoto Univ. 20, 263-289
     (1980).

59   Simon B. : Functional integration and quantum physics. New York :
     Acad. Press. 1979.

60   Sjöstrand J. : On the eigenvalues of a class of hypoelliptic ope-
     rators IV . Ann. Inst. Fourier XXX, 109-169 (1980).

61   Stroock D. : The Malliavin calculus and its applications to second
     order parabolic differential equations. Math. Systems Theory, Part I :
     14, 25-65 (1981). Part II : 14, 141-171 (1981).

62   Stroock D. : The Malliavin calculus : a functional analytic approach.
     J. of Funct. Anal. 44, 212-257 (1981).

63   Stroock D. : Some applications of stochastic calculus to partial
     differential equations. Ecole de probabilités de Saint Flour. Lecture
     Notes in Math. n° 976, pp. 267-382. Berlin : Springer 1983.

64   Stroock D., Varadhan S.R.S. : Multidimensional Diffusion processes. Grundlehren Math. Wissenschaften Band 233, Berlin : Springer 1979.

65   Taniguchi S. : Malliavin's stochastic calculus of variations for manifold valued Wiener functionals and its applications. To appear.

66   Treves F. : Introduction to pseudodifferential operators and Fourier Integral operators. Vol. 1 and 2. New York : Plenum Press 1980.

67   Truman A. : The polygonal path formulation of the Feynman path integral. Lecture Notes in Physics n° 106, S. Albeverio and al. ed. pp. 73-102, Berlin : Springer 1979.

68   Truman A. : Classical mechanics , the diffusion heat equation, and the Schrödinger equation . J. Math Phys. 18, 2308-2315 (1977).

69   Varadhan S.R.S. : Asymptotic probabilities and differential equations. Commun. Pure and Appl. Math., XIX, 261-286 (1966).

70   Varadhan S.R.S. : Diffusion processes in a small time interval. Commun. Pure and Appl. Math. 20, 659-685 (1967).

71   Hida T. : Brownian motion. Applications of Math N° 11, Berlin : Springer 1980.

72   Reed M., Simon B., : Methods of modern mathematical physics. Vol. 1 and 2.   New York : Acad. Press 1972 and 1975.

73   Schwartz L. : Semi-martingales sur des variétés et martingales conformes sur des variétés analytiques complexes. Lecture Notes in Math n° 780. Berlin : Springer 1980.

74 Minakshishundaram S., Pleijel A. : Some properties of the eigen functions of the Laplace operator on a Riemannian manifold. Can. J. Math. 1, 242-256 (1949).

75 Chaleyat - Maurel M., and Jeulin T : Grossissement Gaussien de la filtrations Brownienne. To appear.

76 Seeley R.T. : Complex power of an elliptic operator. Proc. Symp. A.M.S. 10, 288-307 (1967).

77 Ventcell D., Freidlin M.I. : On small random perturbations of dynamical systems. Russian Math. Surveys 25, 1-55 (1970).

78 Yor M. : Remarques sur une formule de P. Lévy. In Séminaire de Probabilités n° XIV. pp. 343-346. Lecture Notes in Math. n° 784, Berlin : Springer 1980.

79 Erdelyi A. : Asymptotic expansions. New York : Dover Pub. 1956.

80 Fadeev L.D., Slavnov A.A. : Gauge fields. Introduction to quantum Theory. London : Benjamin Cummings 1980.

81 Langouche F., Rockaerts D., Tirapegui E. : Functional Integration and semi-classical expansions. Dordrecht : Reidel 1982.

82 Garsia A.M., Rodemich E., Rumsey H. : A real variable lemma and the continuity of paths of some gaussian processes. Indiana Univ. Math. J. 20, 565-578 (1970).

83 Bismut J.M. : Calcul des variations stochastiques et grandes déviations. CRAS 296, 1009-1012 (1983).

84 Bismut J.M. : Le Théorème d'Atiyah-Singer pour les opérateurs elliptiques classiques. C.R.A.S. Série I. To appear (1983).

85 Bismut J.M. : The Atiyah-Singer theorems for classical elliptic operators : a probabilistic approach. To appear in J. Funct. Anal. (1984).

86 Kifer Y. : On the asymptotics of the transition density of processes with small diffusion. Theory of Prob. and Appl. 21, 512-522 (1976).

87 Malliavin P. : Implicit functions in finite corank on the Wiener Space. Proceedings of the Katata Conference in Probability (1981). To appear.

Progress in Mathematics
Edited by J. Coates and S. Helgason

Progress in Physics
Edited by A. Jaffe and D. Ruelle

- A collection of research-oriented monographs, reports, notes arising from lectures or seminars
- Quickly published concurrent with research
- Easily accessible through international distribution facilities
- Reasonably priced
- Reporting research developments combining original results with an expository treatment of the particular subject area
- A contribution to the international scientific community: for colleagues and for graduate students who are seeking current information and directions in their graduate and post-graduate work.

Manuscripts

Manuscripts should be no less than 100 and preferably no more than 500 pages in length.

They are reproduced by a photographic process and therefore must be typed with extreme care. Symbols not on the typewriter should be inserted by hand in indelible black ink. Corrections to the typescript should be made by pasting in the new text or painting out errors with white correction fluid.

The typescript is reduced slightly (75%) in size during reproduction; best results will not be obtained unless the text on any one page is kept within the overall limit of 6x9½ in (16x24 cm). On request, the publisher will supply special paper with the typing area outlined.

Manuscripts should be sent to the editors or directly to: Birkhäuser Boston, Inc., P.O. Box 2007, Cambridge, Massachusetts 02139

PROGRESS IN MATHEMATICS
Already published

PM 1   Quadratic Forms in Infinite-Dimensional Vector Spaces
Herbert Gross
ISBN 3-7643-1111-8, 432 pages, paperback

PM 2   Singularités des systèmes différentiels de Gauss-Manin
Frédéric Pham
ISBN 3-7643-3002-3, 346 pages, paperback

PM 3   Vector Bundles on Complex Projective Spaces
C. Okonek, M. Schneider, H. Spindler
ISBN 3-7643-3000-7, 396 pages, paperback

PM 4   Complex Approximation, Proceedings, Quebec, Canada, July 3-8, 1978
Edited by Bernard Aupetit
ISBN 3-7643-3004-X, 128 pages, paperback

PM 5   The Radon Transform
Sigurdur Helgason
ISBN 3-7643-3006-6, 207 pages, hardcover

PM 6   The Weil Representation, Maslov Index and Theta Series
Gérard Lion, Michèle Vergne
ISBN 3-7643-3007-4, 348 pages, paperback

PM 7   Vector Bundles and Differential Equations
Proceedings, Nice, France, June 12-17, 1979
Edited by André Hirschowitz
ISBN 3-7643-3022-8, 256 pages, paperback

PM 8   Dynamical Systems, C.I.M.E. Lectures, Bressanone, Italy, June 1978
John Guckenheimer, Jürgen Moser, Sheldon E. Newhouse
ISBN 3-7643-3024-4, 305 pages, hardcover

PM 9   Linear Algebraic Groups
T. A. Springer
ISBN 3-7643-3029-5, 314 pages, hardcover

PM 10   Ergodic Theory and Dynamical Systems I
A. Katok
ISBN 3-7643-3036-8, 346 pages, hardcover

PM 11   18th Scandinavian Congress of Mathematicians, Aarhus, Denmark, 1980
Edited by Erik Balslev
ISBN 3-7643-3040-6, 526 pages, hardcover

PM 12   Séminaire de Théorie des Nombres, Paris 1979-80
Edited by Marie-José Bertin
ISBN 3-7643-3035-X, 404 pages, hardcover

PM 13   Topics in Harmonic Analysis on Homogeneous Spaces
Sigurdur Helgason
ISBN 3-7643-3051-1, 152 pages, hardcover

PM 14 Manifolds and Lie Groups, Papers in Honor of Yozô Matsushima
Edited by J. Hano, A. Marimoto, S. Murakami, K. Okamoto, and H. Ozeki
ISBN 3-7643-3053-8, 476 pages, hardcover

PM 15 Representations of Real Reductive Lie Groups
David A. Vogan, Jr.
ISBN 3-7643-3037-6, 776 pages, hardcover

PM 16 Rational Homotopy Theory and Differential Forms
Phillip A. Griffiths, John W. Morgan
ISBN 3-7643-3041-4, 258 pages, hardcover

PM 17 Triangular Products of Group Representations and their Applications
S. M. Vovsi
ISBN 3-7643-3062-7, 142 pages, hardcover

PM 18 Géométrie Analytique Rigide et Applications
Jean Fresnel, Marius van der Put
ISBN 3-7643-3069-4, 232 pages, hardcover

PM 19 Periods of Hilbert Modular Surfaces
Takayuki Oda
ISBN 3-7643-3084-8, 144 pages, hardcover

PM 20 Arithmetic on Modular Curves
Glenn Stevens
ISBN 3-7643-3088-0, 236 pagers, hardcover

PM 21 Ergodic Theory and Dynamical Systems II
A. Katok, editor
ISBN 3-7643-3096-1, 226 pages, hardcover

PM 22 Séminaire de Théorie des Nombres, Paris 1980-81
Marie-José Bertin, editor
ISBN 3-7643-3066-X, 374 pages, hardcover

PM 23 Adeles and Algebraic Groups
A. Weil
ISBN 3-7643-3092-9, 138 pages, hardcover

PM 24 Enumerative Geometry and Classical Algebraic Geometry
Patrick Le Barz, Yves Hervier, editors
ISBN 3-7643-3106-2, 260 pages, hardcover

PM 25 Exterior Differential Systems and the Calculus of Variations
Phillip A. Griffiths
ISBN 3-7643-3103-8, 349 pages, hardcover

PM 26 Number Theory Related to Fermat's Last Theorem
Neal Koblitz, editor
ISBN 3-7643-3104-6, 376 pages, hardcover

PM 27 Differential Geometric Control Theory
Roger W. Brockett, Richard S. Millman, Hector J. Sussmann, editors
ISBN 3-7643-3091-0, 349 pages, hardcover

PM 28 Tata Lectures on Theta I
David Mumford
ISBN 3-7643-3109-7, 254 pages, hardcover

PM 29	Birational Geometry of Degenerations *Robert Friedman and David R. Morrison, editors* ISBN 3-7643-3111-9, 410 pages, hardcover
PM 30	CR Submanifolds of Kaehlerian and Sasakian Manifolds *Kentaro Yano, Masahiro Kon* ISBN 3-7643-3119-4, 223 pages, hardcover
PM 31	Approximations Diophantiennes et Nombres Transcendants *D. Bertrand and M. Waldschmidt, editors* ISBN 3-7643-3120-8, 349 pages, hardcover
PM 32	Differential Geometry *Robert Brooks, Alfred Gray, Bruce L. Reinhart, editors* ISBN 3-7643-3134-8, 267 pages, hardcover
PM 33	Uniqueness and Non-Uniqueness in the Cauchy Problem *Claude Zuily* ISBN 3-7643-3121-6, 185 pages, hardcover
PM 34	Systems of Microdifferential Equations *Masaki Kashiwara* ISBN 0-8176-3138-0 ISBN 3-7643-3138-0, 182 pages, hardcover
PM 35	Arithmetic and Geometry Papers Dedicated to I. R. Shafarevich on the Occasion of His Sixtieth Birthday Volume I Arithmetic *Michael Artin, John Tate, editors* ISBN 3-7643-3132-1, 373 pages, hardcover
PM 36	Arithmetic and Geometry Papers Dedicated to I. R. Shafarevich on the Occasion of His Sixtieth Birthday Volume II Geometry *Michael Artin, John Tate, editors* ISBN 3-7643-3133-X, 495 pages, hardcover
PM 37	Mathématique et Physique *Louis Boutet de Monvel, Adrien Douady, Jean-Louis Verdier, editors* ISBN 0-8176-3154-2 ISBN 3-7643-3154-2, 454 pages, hardcover
PM 38	Séminaire de Théorie des Nombres, Paris 1981-82 *Marie-José Bertin, editor* ISBN 0-8176-3155-0 ISBN 3-7643-3155-0, 359 pages, hardcover
PM 39	Classical Algebraic and Analytic Manifolds *Kenji Ueno, editor* ISBN 0-8176-3137-2 ISBN 3-7643-3137-2, 644 pages, hardcover
PM 40	Representation Theory of Reductive Groups *P. C. Trombi, editor* ISBN 0-8176-3135-6 ISBN 3-7643-3135-6, 308 pages, hardcover
PM 41	Combinatorics and Commutative Algebra *Richard P. Stanley* ISBN 0-8176-3112-7 ISBN 3-7643-3112-7, 102 pages, hardcover
PM 42	Théorèmes de Bertini et Applications *Jean-Pierre Jouanolou* ISBN 0-8176-3164-X ISBN 3-7643-3164-X, 140 pages, hardcover

PM 43   Tata Lectures on Theta II
         David Mumford
         ISBN 0-8176-3110-0
         ISBN 3-7643-3110-0, 272 pages, hardcover

PM 44   Infinite Dimensional Lie Algebras
         Victor G. Kac
         ISBN 0-8176-3118-6
         ISBN 3-7643-3118-6, 245 pages, hardcover

PROGRESS IN PHYSICS
Already published

PPh 1 Iterated Maps on the Interval as Dynamical Systems
       Pierre Collet and Jean-Pierre Eckmann
       ISBN 3-7643-3026-O, 256 pages, hardcover

PPh 2 Vortices and Monopoles, Structure of Static Gauge Theories
       Arthur Jaffe and Clifford Taubes
       ISBN 3-7643-3025-2, 294 pages, hardcover

PPh 3 Mathematics and Physics
       Yu. I. Manin
       ISBN 3-7643-3027-9, 112 pages, hardcover

PPh 4 Lectures on Lepton Nucleon Scattering and Quantum Chromodynamics
       W. B. Atwood, J. D. Bjorken, S. J. Brodsky, and R. Stroynowski
       ISBN 3-7643-3079-1, 574 pages, hardcover

PPh 5 Gauge Theories: Fundamental Interactions and Rigorous Results
       P. Dita, V. Georgescu, R. Purice, editors
       ISBN 3-7643-3095-3, 406 pages, hardcover

PPh 6 Third Workshop on Grand Unification
       University of North Carolina, Chapel Hill, April 15-17, 1982
       P. H. Frampton, S. L. Glashow, and H. van Dam, editors
       ISBN 3-7643-3105-4, 382 pages, hardcover

PPh 7 Scaling and Self-Similarity in Physics
       (Renormalization in Statistical Mechanics and Dynamics)
       J. Fröhlich, editor
       ISBN 3-7643-3168-2
       ISBN 0-8176-3168-2, 440 pages, hardcover

PPh 8 Workshop on Non-Perturbative Quantum Chromodynamics
       K. A. Milton, M. A. Samuel, editors
       ISBN 3-7643-3127-5
       ISBN 0-8176-3127-5, 284 pages, hardcover

PPh 9 Fourth Workshop on Grand Unification
       University of Pennsylvania, Philadelphia, April 21-23, 1983
       H. A. Weldon, P. Langacker, P. J. Steinhardt, editors
       ISBN 3-7643-3127-5
       ISBN 0-8176-3127-5, 284 pages, hardcover